Springer Theses

Recognizing Outstanding Ph.D. Research

Aims and Scope

The series "Springer Theses" brings together a selection of the very best Ph.D. theses from around the world and across the physical sciences. Nominated and endorsed by two recognized specialists, each published volume has been selected for its scientific excellence and the high impact of its contents for the pertinent field of research. For greater accessibility to non-specialists, the published versions include an extended introduction, as well as a foreword by the student's supervisor explaining the special relevance of the work for the field. As a whole, the series will provide a valuable resource both for newcomers to the research fields described, and for other scientists seeking detailed background information on special questions. Finally, it provides an accredited documentation of the valuable contributions made by today's younger generation of scientists.

Theses are accepted into the series by invited nomination only and must fulfill all of the following criteria

- They must be written in good English.
- The topic should fall within the confines of Chemistry, Physics, Earth Sciences, Engineering and related interdisciplinary fields such as Materials, Nanoscience, Chemical Engineering, Complex Systems and Biophysics.
- The work reported in the thesis must represent a significant scientific advance.
- If the thesis includes previously published material, permission to reproduce this must be gained from the respective copyright holder.
- They must have been examined and passed during the 12 months prior to nomination.
- Each thesis should include a foreword by the supervisor outlining the significance of its content.
- The theses should have a clearly defined structure including an introduction accessible to scientists not expert in that particular field.

More information about this series at http://www.springer.com/series/8790

Shuzhe Shi

Soft and Hard Probes of QCD Topological Structures in Relativistic Heavy-Ion Collisions

Doctoral Thesis accepted by Indiana University,
Bloomington, IN, USA

 Springer

Shuzhe Shi
Physics Department
McGill University
Montréal, QC, Canada

ISSN 2190-5053   ISSN 2190-5061   (electronic)
Springer Theses
ISBN 978-3-030-25484-1   ISBN 978-3-030-25482-7   (eBook)
https://doi.org/10.1007/978-3-030-25482-7

This Springer imprint is published by the registered company Springer Nature Switzerland AG.
The registered company address is: Gewerbestrasse 11, 6330 Cham, Switzerland

Supervisor's Foreword

Nuclear physics studies the subatomic structures and the associated strong interactions that are at the heart of all visible matter. The basic theory, quantum chromodynamics (QCD), is however difficult to solve due to its strongly coupled nature. Even its fundamental degrees of freedom, quarks and gluons, are deeply bound inside hadrons under normal circumstances, a phenomenon called "confinement." By colliding heavy ions at high energies, physicists are able to "break up" hadrons and create a hot "subatomic soup" directly of quarks and gluons—a new form of matter called a quark-gluon plasma (QGP). The QGP forms at a temperature of one trillion degrees or higher and briefly occupies the baby universe. This exotic matter holds the key for understanding our cosmic origin and has now been re-created in laboratories at the Relativistic Heavy Ion Collider (RHIC) and the Large Hadron Collider (LHC), allowing the study of QCD properties in a quark-gluon many-body system. The investigation of the unusual properties and novel emergent phenomena in such an "extreme matter" is a major research frontier in today's nuclear science.

It has been long believed that most of the highly nontrivial and non-perturbative dynamics of QCD has to do with the emergence of gluon topological configurations that arise from "twisting" gluon fields across space-time. While calculable on first-principle lattice QCD approach, such gluon fields are however difficult to detect experimentally. In addition, the roles of topological component in the quark-gluon plasma and their manifestation in heavy ion collisions have been intriguing yet unclear. The thesis work of Dr. Shuzhe Shi has precisely addressed such important questions and significantly advanced our understanding.

One important fact about the gluon topology is that it can be manifested through light quarks when they couple together. The changes in gluon topology are always accompanied by the exact amount of corresponding changes in quark chirality, through the famous chiral anomaly. The first part of Dr. Shi's thesis has investigated the soft probe of strong interaction topological fluctuations in the quark-gluon plasma (QGP) via a novel phenomenon related to quark chirality: the Chiral Magnetic Effect (CME). In this part, Dr. Shi has made instrumental contributions in establishing the first comprehensive tool: a fluid dynamical realization of CME which we call Anomalous-Viscous Fluid Dynamics (AVFD). The AVFD has demonstrated CME as a quantitative explanation of relevant experimental data and also made testable predictions for upcoming analyses of the isobaric collisions at RHIC.

The presence of topological component in QGP can also be probed by the "high-speed bullets," or jets, penetrating through the plasma. Their behaviors sensitively depend upon the scatterings they experience with the constituents and thus the energy they lose in the hot medium. Such jet energy loss provides unique "tomography" about what are inside the plasma. The second part of this thesis deals with the hard probe of strongly coupled quark-gluon plasma. In particular, the study addresses the basic question on the non-perturbative color structure in the QGP via jet energy loss observables. Dr. Shi has significantly advanced the so-called CUJET/CIBJET framework and used it to disclose the topological degrees of freedom in quark-gluon plasma. A major outcome is a successful description of a large body of relevant experimental data at RHIC and LHC, along with predictions for XeXe collisions and a few heavy flavor energy loss observables. A key finding on the QGP color structure has shown that a chromo-magnetic topological component is indispensable.

Each component of these studies is highly challenging, and Dr. Shi has addressed them with unparalleled dedication and innovative approaches. The research of this thesis has generated widespread impacts in the field of quark-gluon plasma and high-energy nuclear collisions, with many of the key results being used by experimental collaborations at RHIC and LHC for comparison and being highlighted by plenary talks at a series of major conferences in the field. Dr. Shi's thesis represents outstanding scholarly achievements that are of the highest caliber. It is thus a great pleasure that his work has now been recognized by the Springer Thesis Award. This published version of his thesis will serve as an excellent reference and continue to deliver impacts for future research.

Bloomington, IN, USA   Dr. Jinfeng Liao
June 2019

Acknowledgments

First of all, I thank my advisor, Dr. Jinfeng Liao. This dissertation would never have been done without his penetrating intuition and lightening ideas. In the meanwhile, I am sincerely indebted to his continuously strong support, as well as his friendly education on performing researches and presenting the results. I would like to thank Dr. Miklos Gyulassy for his collaboration in the CUJET/CIBJET project. I really appreciate his invaluable guidance on jet physics, as well as his kindness and encouragements to young students like me. Besides, I would like to take this special opportunity to thank my *B.S.* and *M.S.* supervisor at Tsinghua University, Dr. Pengfei Zhuang, who leads me to the field of high-energy nuclear physics and helps me greatly on starting research. I am so obliged to all the suggestions, help, and encouragements.

I would like to thank other members in my Ph.D. candidacy committee, Drs. Charles J. Horowitz, Scott W. Wissink, and Shixiong Zhang, for all their valuable suggestions. I also would like to thank Anping Huang, Yin Jiang, Kangle Li, Elias Lilleskov, Yi Yin, Jiechen Xu, and Hui Zhang for their fruitful collaborations on research related to these dissertation and Xingyu Guo, Ulrich Heinz, Yuji Hirono, Xu-Guang Huang, Dmitri Kharzeev, Roy A. Lacey, Hui Li, Mark Mace, Niseem Magdy, Jacquelyn Noronha-Hostler, Jorge Noronha, Sören Schlichting, Sayantan Sharma, Chun Shen, Fuqiang Wang, Liwen Wen, Gang Wang, Xiaoliang Xia, Nu Xu, and Ho-Ung Yee for their very helpful discussions. These are essential to complete and improve the research in this thesis.

Personally, I would incline to regard my "Ph.D. career" as the combination of two periods, which includes 3 years in Tsinghua University as a M.S. student and another 3 years in IU, Bloomington, officially as an Ph.D. student. I thank all my teachers, group members, and friends during both periods. Especially, I would like to thank other members of the Indiana University Nuclear Theory Center: Charles Horowitz, Emilie Passemar, Farrooh Fattoyev, Zidu Lin, Miguel Lopez Ruiz, and Vicent Mathieu, as well as Zhe Xu, Lianyi He, Gaoqing Cao, Baoyi Chen, Zhengyu Chen, Wei Dai, Bohao Feng, Xingyu Guo, Hang He, Anping Huang, Yunpeng Liu, Minjie Luo, Yin Jiang, Shijun Mao, Lingxiao Wang, Ziyue Wang, Tao Xia, Jiaxing Zhao, and Kai Zhou from Tsinghua University HEP group. The delightful discussions with you guys had greatly benefitted my research and my knowledge of physics. It is my great honor to meet everyone of you.

There is still a long list of friends I need to thank. In particular, I would like to thank Hao Ding, Dmitri Kalinkin, Zonghao Li, Huanzhao Liu, Shufan Lu, Jianchun Yin; the "Bloomington Hot-Pot Group"—Peilian Li, Ting Lin, Zidu Lin, Zhi Liu, Jiazhou Shen, Qinghui Sun, Miao Wang, Yuanheng Xie, Wencao Yang, Hui Zhang; as well as my friends at Tsinghua—Ang Gao, Xingyu Guo, Jianhong Hu, Jiying Jia, Tian Lan, Qingkai Qian, Chenyang Tang, Lu Tian, Dayou Yang, Zhe Yang, Qibo Zeng, Cheng Zhao, and Wei Zhao. I really treasure the wonderful days we spent together, as well as our conversations inside and outside physics, which broaden my knowledge and, more importantly, allow me to enjoy the colorful life!

Most importantly, I would like to express my deepest sense of gratitude to my family, especially my parents. I realized that it has been 10 years since I left home. Your continuous endless love and encouragement make me always feel your company—I have never walked alone. Thank You!

Last but not the least, I am grateful to the financial support by the US Department of Energy, Office of Science and of Nuclear Physics, within the framework of the Beam Energy Scan Theory (BEST) Topical Collaboration, and by the National Science Foundation under Grant No. PHY-1352368. I also would like to acknowledge IU's Big Red II and Karst clusters, which generously provided $\sim 3 \times 10^6$ cpu h ($\sim 3 \times 10^2$ cpu years) free computational time. These clusters are supported in part by Lilly Endowment, Inc., through its support for the Indiana University Pervasive Technology Institute and in part by the Indiana METACyt Initiative. The Indiana METACyt Initiative at IU was also supported in part by Lilly Endowment, Inc.

Montreal, QC, Canada   Shuzhe Shi
June 2019

Contents

Parts of This Thesis Have Been Published in the Following Journal Articles

1. S. Shi, Y. Jiang, E. Lilleskov and J. Liao, *"Anomalous Chiral Transport in Heavy Ion Collisions from Anomalous-Viscous Fluid Dynamics"*, Annals Phys. **349** (2018) 50–72. https://doi.org/10.1016/j.aop.2018.04.026
2. Y. Jiang, S. Shi, Y. Yin and J. Liao, *"Quantifying Chiral Magnetic Effect from Anomalous-Viscous Fluid Dynamics"*, Chin. Phys. **C 42** (2018) 1, 011001. https://doi.org/10.1088/1674-1137/42/1/011001
3. A. Huang, S. Shi, Y. Jiang, J. Liao and P. Zhuang, *"Complete and Consistent Chiral Transport from Wigner Function Formalism"*, Phys. Rev. **D 98** (2018) 036010. https://doi.org/10.1103/PhysRevD.98.036010
4. A. Huang, Y. Jiang, S. Shi, J. Liao and P. Zhuang, *"Out-of-Equilibrium Chiral Magnetic Effect from Chiral Kinetic Theory"*, Phys. Lett. **B 777** (2018) 177. https://doi.org/10.1016/j.physletb.2017.12.025
5. S. Shi, K. Li, and J. Liao, *"Searching for the Subatomic Swirls in the CuCu and CuAu Collisions"*, Phys. Lett. **B 788** (2019) 409. https://doi.org/10.1016/j.physletb.2018.09.066
6. S. Shi, J. Liao and M. Gyulassy, *"Global Constraints from RHIC and LHC on Transport Properties of QCD Fluid with the CUJET/CIBJET Framework"*, Chin. Phys. **C 43** (2019) 4, 044101. https://doi.org/10.1088/1674-1137/43/4/044101
7. S. Shi, J. Liao and M. Gyulassy, *"Probing the Color Structure of the Perfect QCD Fluids via Soft-Hard-Event-by-Event Azimuthal Correlations"*, Chin. Phys. **C 42** (2018) 10, 104104. https://doi.org/10.1088/1674-1137/42/10/104104

Chapter 1
Introduction

In relativistic heavy-ion collision experiments, a new phase of matter—the QCD Plasma—is created. In such QCD Plasma the color degrees of freedom, e.g., quarks and gluons, are deconfined and the chiral symmetry is restored. This dissertation focuses on both soft (particles with transverse momentum $p_T < 2\,\mathrm{GeV}$) and hard probes ($p_T > 10\,\mathrm{GeV}$) of the QCD Plasma, especially its topological properties, by performing quantification study of the Chiral Magnetic Effect (CME) which reflects the topological charge transition, as well as jet quenching observables as a probe of the chromo-magnetic degrees of freedom.

On one hand, we perform a quantitative, precise study of the CME, which is a macroscopic manifestation of fundamental chiral anomaly in a many-body system of chiral fermions, and emerges as anomalous transport current in the fluid dynamics framework. We develop the Anomalous-Viscous Fluid Dynamics (AVFD) framework, which implements the anomalous fluid dynamics to describe the evolution of fermion currents in QGP, on top of the neutral bulk background described by the VISH2+1 hydrodynamic simulations for heavy-ion collisions. With this new tool, we systematically investigate the dependence of the CME signal to a series of theoretical inputs and associated uncertainties. With realistic estimates of initial conditions and magnetic field lifetime, the predicted CME signal is quantitatively consistent with measured charge separation data in 200 GeV Au–Au collisions. Based on both event-averaged and event-by-event analyses of Au–Au collisions, we further make predictions for the CME observable in the on-going isobaric (Ru–Ru v.s. Zr–Zr) collisions, which could provide the most decisive test of the CME in heavy-ion collisions.

On the other hand, a systematic analysis of jet quenching observables is performed, based on the CUJET3.1 and CIBJET simulation frameworks. The CUJET3.1 framework consistently combines viscous hydrodynamic fields predicted by smooth VISHNU2+1 and the DGLV jet energy loss theory generalized to sQGMP fluids with color structure including both electric and magnetic components. Based on a global, quantitative comparison with a comprehensive set of

© Springer Nature Switzerland AG 2019
S. Shi, *Soft and Hard Probes of QCD Topological Structures in Relativistic Heavy-Ion Collisions*, Springer Theses,
https://doi.org/10.1007/978-3-030-25482-7_1

experimental data, we find that recent correlation data favor a temperature dependent color composition including bleached chromo-electric components, i.e., quark and gluon, and an emergent chromo-magnetic degrees of freedom. Furthermore, a comprehensive dynamical framework—CIBJET—is developed to calculate, on an event-by-event basis, the dependence of correlations between soft and hard azimuthal flow angle harmonics on the color composition of near-perfect QCD fluids. CIBJET combines consistently predictions of event-by-event VISHNU2+1 viscous hydrodynamic fluid fields with CUJET3.1 predictions of event-by-event jet quenching. We find event-by-event fluctuation of the bulk background creates a non-vanishing v_3, while it does not effect R_{AA} and has only limited influence on v_2.

1.1 Strong Interaction and Quantum ChromoDynamics

Symmetry principles play instrumental roles in the construction of our most basic physical theories. It means the invariance of physical laws under transformations, such as space-time translation, reflection, rotation, boost, etc. In particular, symmetry principles are essential in the development of nuclear physics and particle physics, which deal with systems with small size or/and high energy scale, hence both quantum effect and relativistic correction shall be considered. To study such systems, theoretical physicists invented quantum field theory (QFT), a quantum theory being invariant under spatial rotation and Lorentz boost.

The first success of QFT comes from Quantum ElectroDynamics (QED) [1–4] by its highly accurate description of phenomena dominated by electromagnetic interaction. QED is a theory coupling spin-$\frac{1}{2}$ charged fermions with spin-1 photons, with the former described by Dirac spinors ψ, while the latter described by Abelian vector fields A^μ. The Lagrangian density of QED is expressed as

$$\mathcal{L}_{\text{QED}} = \mathcal{L}_{\text{Maxwell}} + \mathcal{L}_{\text{Dirac}} - q\overline{\psi}(\gamma^\mu A_\mu)\psi, \qquad (1.1)$$

$$\mathcal{L}_{\text{Dirac}} = \overline{\psi}(i\gamma^\mu \partial_\mu - m)\psi, \qquad (1.2)$$

$$\mathcal{L}_{\text{Maxwell}} = -\frac{1}{4}F^{\mu\nu}F_{\mu\nu}, \qquad (1.3)$$

where the field tensor, $F_{\mu\nu} = \partial_\mu A_\nu - \partial_\nu A_\mu$, is defined in the same way as classical electrodynamics. Benefited from its small coupling constant, one can apply perturbation expansion to QED calculations and provide extremely accurate predictions of observables such as anomalous magnetic moment of the electron [2] and the Lamb shift of the energy levels of hydrogen [5].

QED is a relativistic theory respecting Lorentz invariance. In the meanwhile, it is also invariant under the gauge transformation—a local phase shift of the Dirac spinor as well as corresponding change of the Maxwell vector field:

$$\psi \to e^{ie\alpha(x)}\psi, \quad A_\mu(x) \to A_\mu(x) - \partial_\mu \alpha(x) = e^{ie\alpha(x)}\left[A_\mu(x) + \frac{1}{ie}\partial_\mu\right]e^{-ie\alpha(x)}.$$

$$(1.4)$$

It is straightforward to verify the invariance of field strength tensor, Maxwell Lagrangian as well as QED Lagrangian:

$$F_{\mu\nu} \to F_{\mu\nu}, \qquad \mathcal{L}_{\text{Maxwell}} \to \mathcal{L}_{\text{Maxwell}}, \qquad \mathcal{L}_{\text{QED}} \to \mathcal{L}_{\text{QED}}. \qquad (1.5)$$

Hence, a Maxwell field (1.3) is also referred to as Abelian gauge field, where Abelian means its trivial commutation relation:

$$[A^\mu, A^\nu] \equiv A^\mu A^\nu - A^\nu A^\mu = 0. \qquad (1.6)$$

Based on QED, Yang and Mills further generalize the theory from a single-component field to multiple-component fields [6]. In the latter case, the gauge transformation is a local SU(N) transformation causing not only phase shift but also rotation in component-index space:

$$\mathbf{U} \equiv \exp[ig\boldsymbol{\alpha}(x)] = \exp\left[ig\alpha^a(x)\frac{\lambda^a}{2}\right], \qquad (1.7)$$

$$\boldsymbol{\psi} \to \mathbf{U} \cdot \boldsymbol{\psi}, \qquad \mathbf{A}_\mu(x) \to \mathbf{U} \cdot \left[\mathbf{A}_\mu(x) + \frac{1}{ig}\partial_\mu\right] \cdot \mathbf{U}^\dagger. \qquad (1.8)$$

It is worth noting that a set consisting of all $\mathbf{U}$'s forms *Lie group*, with its generators λ^a form a *Lie algebra*. The Lie algebra is defined through commutation relations:

$$\left[\frac{\lambda^a}{2}, \frac{\lambda^b}{2}\right] = if^{abc}\frac{\lambda^c}{2}, \qquad (1.9)$$

where f^{abc} are called *structure constants*. Nonzero structure constants lead to *non-Abelianity* of the coupling gauge fields. One can represent the (non-Abelian) gauge field as

$$\mathbf{A}_\mu(x) = A_\mu^a(x)\frac{\lambda^a}{2}, \qquad (1.10)$$

and further construct corresponding field strength tensor as well as Lagrangian density

$$G_{\mu\nu}^a = \partial_\mu A_\nu^a - \partial_\nu A_\mu^a - gf^{abc}A_\mu^b A_\nu^c, \qquad (1.11)$$

$$\mathcal{L}_{\text{YM}} = -\frac{1}{4}G^{a,\mu\nu}G_{\mu\nu}^a. \qquad (1.12)$$

The field theory of strong interaction, named as Quantum ChromoDynamics (QCD), is a theory respecting SU(3) gauge symmetry. It couples eight species of vector fields (*gluons*) with three species (also called *colors*) of spinor fields (*quarks*), while the eight generators λ^a are 3×3 matrices often taking the form of *Gell-Mann matrices* [7].[1] The QCD Lagrangian is constructed as

$$\mathcal{L}_{QCD} = \mathcal{L}_{YM} + \overline{\psi} \cdot (i\gamma^{\mu}\mathbf{D}_{\mu} - m) \cdot \psi \tag{1.13}$$

where the covariant derivative is defined as

$$(D_{\mu}\psi)^i = \left[\delta^{ij}\partial_{\mu} + igA_{\mu}^a\left(\frac{\lambda^a}{2}\right)^{ij}\right]\psi^j \tag{1.14}$$

with indices $a, b, c = 1, \ldots, 8$, while $i, j = 1, 2, 3$. The nature of being non-Abelian as well as strongly coupled brings a lot of fascinating phenomena in QCD predominating systems, such as asymptotic freedom and color confinement.

Asymptotic Freedom The renormalization of QCD predicts a drastically different form of running coupling from QED, as a consequence of the gluon self-interactions. For interactions at energy scale Q, the strong coupling $\alpha_s \equiv g^2/(4\pi)$ reads ([8, 9], G. 't Hooft, unpublished, 1972):

$$\alpha_s(Q^2) = \frac{4\pi}{(11 - 2N_f/3)\ln(Q^2/\Lambda_{QCD}^2)}, \tag{1.15}$$

with QCD scaling $\Lambda_{QCD} \sim 200\,\text{MeV}$, while N_f denotes the number of quark flavors. As shown in Fig. 1.1, the coupling α_s decreases as energy scale Q increases, and it reaches the perturbative regime for $Q \gtrsim 5\,\text{GeV}$. For the hard process with high momentum transfers $Q \gg \Lambda_{QCD}$ corresponding to small coupling constant, one can provide theoretical predictions by perturbative QCD calculations. Consequently, experimental observables corresponding to such process, like jet physics in high energy $e^+ + e^-$ and $p + p$ collisions, serve as a crucial test of QCD.

1.1.1 Color Confinement and QCD Phase Transition

In particle physics, the quark model is widely accepted from its great success in explaining hadron mass hierarchy as well as weak interaction phenomena. However, single free quarks are never directly observed or found in experiment. Such phenomenon is attributed as *color confinement*, which indicates that particles carrying nonzero color charge, such as quarks and gluons, cannot be isolated in cold

---

[1]It might be worth mentioning that in this representation, the first three generators $\lambda_1, \lambda_2, \lambda_3$ form a SU(2) sub-algebra of the gauge field theory. Such property is frequently employed in the explicit examples in this chapter.

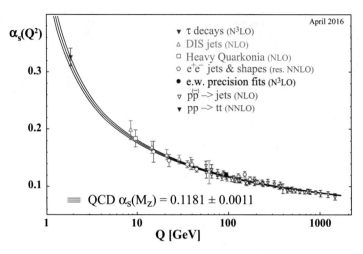

Fig. 1.1 Summary of measurements on running of the strong coupling $\alpha_s(Q^2)$. Reproduced (with permission) from Particle Data Group Booklet [10]

and dilute conditions. There is not yet an analytic proof of color confinement, and remains to be a question attracting numerous efforts.

However, while the color degree of freedom is confined in "normal conditions," it could be liberated in extreme conditions that is at high temperature and/or density. Such transition between color confinement/deconfinement is named as *QCD phase transition*, and the phase structure is schematically shown in Fig. 1.2. For a system at high temperature $T > T_c \sim 160\,\text{MeV} \sim 10^{12}$ K, it creates a new phase of matter, which is usually referred to as *quark–gluon plasma* (QGP), reflecting the liberation of the quark and gluon degrees of freedom. Such hot dense environment is believed to exist in the very early stage of Big Bang, and has been reproduced in ultra-relativistic heavy-ion collision experiments. In such experiments, the existence of the QGP is indicated by many properties of final state particles, such as the quark number scaling of the collective flow [12], quenched jet observables [13], as well as suppressed production rate of J/ψ meson [14, 15]. Presently, these experiments are done at the collider facilities such as Brookhaven National Laboratory's Relativistic Heavy-Ion Collider (RHIC), as well as the CERN's Large Hadron Collider (LHC). While the details of these experiments will be discussed in Sect. 1.3, it might be worth emphasizing here that experiment efforts have been made to study the properties of QGP and especially the characteristics of QCD phase transition. At Brookhaven National Laboratory, the Beam Energy Scan (BES) program [11, 16] is in progress, aiming to study the QCD phase structure, and search for the potential critical point of the phase transition. In the BES program it runs collision experiments for different species of nuclei at different beam energies, produces plasma in different conditions (see yellow spots in Fig. 1.2), and consequently is able to probe different regions of the phase diagram. Recent experimental and theoretical progresses can be found in e.g. [16–20].

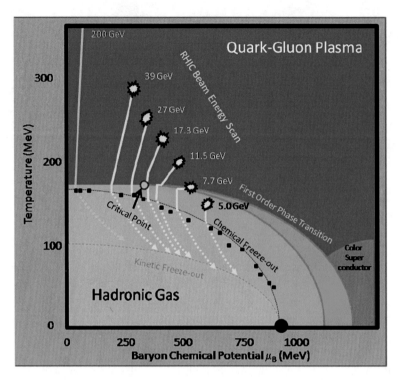

Fig. 1.2 Schematic plot of QCD phase diagram. Figure was reproduced (with permission) from preview of RHIC Beam Energy Scan program [11]

1.2   Topological Properties of Yang–Mills Field

Besides confinement–deconfinement phase transition, the non-Abelian nature also brings nontrivial topological structures of the theory. In particular, in the environment created in relativistic heavy-ion collusions, the particles are strongly coupled and one can see, in the following section, that such topological structures play an important role. In this dissertation, we focus on the topologically nontrivial structure of QCD, and discuss how phenomena in ultra-relativistic heavy-ion collisions would serve as probes of such topological structure. Before discussing about the phenomenological characteristics of QCD, let us first focus on what properties we can find from the classical solutions of Yang–Mills field, namely the solutions to the Equation of Motion (Euler–Lagrange Equation) corresponding to the Yang–Mills Lagrangian (1.12):

$$0 = \partial^\mu \partial_\mu A^{a\nu} - \partial^\nu \partial_\mu A^{a\mu} + g f^{abc} \partial_\mu (A^{b\mu} A^{c\nu})$$
$$+ g f^{abc} A_\mu^b (\partial^\mu A^{c\nu} - \partial^\nu A^{c\mu}) + g^2 f^{abc} f^{cde} A_\mu^b A^{d\mu} A^{e\nu}. \qquad (1.16)$$

Different from Maxwell fields, the non-Abelian nature of Yang–Mills fields causes complexity of solving these equations. One obvious aspect is that the E.o.M. is non-linear with respect to $A(x)$, hence it is not possible to have general/complete solution to these equations. In this section we discuss some special solutions to the Yang–Mills equations, and especially focus on the topological properties.

1.2.1 Degeneracy of QCD Vacua

It is obvious to see that $A^\mu \equiv 0$ ($\mathbf{A}^\mu \equiv \mathbf{0}$) is a trivial solution to the Euler–Lagrange equation of Maxwell (Yang–Mills) field, with corresponding Fermion fields being described by the well-known solution of free Dirac spinors. Then one can take a step forward to find another solution by applying a gauge transformation on such null/vacuum solution, and obtain the *pure gauge* solution:

$$A^{\mathrm{pg}}_\mu(x) = -\partial_\mu \alpha(x) \tag{1.17}$$

for Maxwell field, while

$$\mathbf{A}^{\mathrm{pg}}_\mu(x) = \mathbf{U} \cdot \left[\frac{1}{ig}\partial_\mu\right] \cdot \mathbf{U}^\dagger \tag{1.18}$$

for Yang–Mills field. One can find $A^{\mathrm{pg}}_\mu(x) = -\partial_\mu \alpha(x)$ is again trivial of being the surplus degree of freedom of the electromagnetic field vector. On the contrary, benefited from its non-Abelian nature, the purely gauge solution (1.18) to the Yang–Mills field has nontrivial structures for some specific types of gauge transformations.

Given any arbitrary configuration of Yang–Mills field $\mathbf{A}^\mu(x)$, one can define Chern–Simons current [21, 22]

$$K^\mu = \epsilon^{\mu\nu\rho\sigma} A_{a\nu}\left[G_{a\rho\sigma} - \frac{g}{3}f_{abc}A_{b\rho}A_{c\sigma}\right], \tag{1.19}$$

as well as the Chern–Simons number

$$N_{CS} = \frac{g^2}{16\pi^2}\int K^0 \mathrm{d}^3 x = \frac{g^2}{16\pi^2}\int \mathrm{d}^3 x\, \epsilon^{ijk}\left(A^a_i \partial_j A^a_k - \frac{g}{3}f^{abc}A^a_i A^b_j A^c_k\right). \tag{1.20}$$

It is worth mentioning that for pure gauge solutions, the Chern–Simons number is equivalent to the winding number of the gauge transformation:

$$n_w = \frac{1}{24\pi^2}\int \mathrm{d}^3 x\, \epsilon^{ijk}\mathrm{Tr}\left[(\mathbf{U}\cdot\partial_i\mathbf{U}^\dagger)\cdot(\mathbf{U}\cdot\partial_j\mathbf{U}^\dagger)\cdot(\mathbf{U}\cdot\partial_k\mathbf{U}^\dagger)\right]. \tag{1.21}$$

To see the topologically nontrivial structure of QCD vacua, let us take an explicit example. As a naive generalization from the SU(2) Yang–Mills theory, one can write down a gauge transformation

$$\mathbf{U} = \exp\left[-i\frac{n\pi(x\lambda^1 + y\lambda^2 + z\lambda^3)}{\sqrt{\rho^2 + x^2 + y^2 + z^2}}\right],\qquad (1.22)$$

where ρ is a parameter characterizing the size scale. Such transformation has winding number $n_w = n$, hence yields topologically nontrivial pure gauge Yang–Mills field with $N_{CS} = n$. Even though, such $\mathbf{A}_\mu^{\mathrm{pg}}$'s are associated with vanishing field strength $\mathbf{G}_{\mu\nu} \equiv 0$ and consequently remain to be vacuum solutions. Thus, we obtain a series of vacuum states, called *QCD vacua*, which have topologically nontrivial structure $N_{CS} = \ldots, -2, -1, 0, +1, +2, \ldots$. On the other hand, such vacua solutions are also interesting in the way that they provide boundary conditions to construct other topologically nontrivial structures, due to their vanishing energy density near boundary, which is a prerequisite of the convergence of the system energy.

1.2.2 Instanton and Sphaleron

Driven by the fact that QCD vacua are degenerate, one natural question would be: whether there could be a state linking two QCD vacua. This can be seen by taking the Euclidian space, with metric $(+, +, +, +)$, in which case the Yang–Mills action

$$\begin{aligned}
S_{\mathrm{YM}} &= -\frac{1}{4}\int d^4x\, G_{\mu\nu}^a G^{a\mu\nu} = -\frac{1}{8}\int d^4x\left[G_{\mu\nu}^a G^{a\mu\nu} + \widetilde{G}_{\mu\nu}^a \widetilde{G}^{a\mu\nu}\right]\\
&= \frac{1}{8}\int d^4x\left[\left(G_{\mu\nu}^a \pm \widetilde{G}_{\mu\nu}^a\right)^2 \mp 2G_{\mu\nu}^a \widetilde{G}^{a\mu\nu}\right]\\
&\geq \mp\frac{1}{4}\int d^4x\left(G_{\mu\nu}^a \widetilde{G}^{a\mu\nu}\right)
\end{aligned}\qquad (1.23)$$

reaches its minimum when $G_{\mu\nu}^a = (\pm)\widetilde{G}_{\mu\nu}^a$, i.e., the gluonic stress tensor is self-dual or anti-self-dual in Euclidean space. It is worth mentioning that the Bianchi equality $[\mathbf{D}^\mu, \widetilde{\mathbf{G}}_{\mu\nu}^a] \equiv 0$ ensures such solution also satisfies the Yang–Mills equations. Also from the Yang–Mills equations, one can prove that

$$G^{a\mu\nu}\widetilde{G}_{\mu\nu}^a = \partial_\mu K^\mu.\qquad (1.24)$$

Assuming that the Chern–Simons current K^μ vanishes at spatial infinity, the minimum of S_{YM} is associated with the Chern–Simons charge via

$$\frac{1}{4}\int d^4x\, G^a_{\mu\nu}\tilde{G}^{a\mu\nu} = \frac{1}{4}\int K_\mu\, d\sigma^\mu \tag{1.25}$$

$$= \frac{1}{4}\int_{-\infty}^{\infty} dt\,\frac{d}{dt}\int d^3x\, K^0$$

$$= \frac{8\pi^2}{g^2}\left(N_{CS}|_{t=+\infty} - N_{CS}|_{t=-\infty}\right)$$

$$\equiv \frac{8\pi^2}{g^2}\nu,$$

where ν is the quantized topological charge being the difference between two Chern–Simons numbers. Solutions with $\nu = +1\ (-1)$ are called *instantons (anti-instantons)* [23–25]. Again, extended from SU(2) Yang–Mills field, one can write down an instanton solution explicitly as:

$$\mathbf{A}^\mu = \frac{2}{g(r^2+\rho^2)}\left[(x,\tau,-z,y)\frac{\lambda^1}{2} + (y,z,\tau,-x)\frac{\lambda^2}{2} + (z,-y,x,\tau)\frac{\lambda^3}{2}\right], \tag{1.26}$$

where $r \equiv \sqrt{\tau^2+x^2+y^2+z^2}$ corresponds to the proper time, and ρ is a free parameter characterizing the instanton size. One can find that solution (1.26) has Chern–Simons number of $N_{CS} = \frac{\tau(2\tau^2+3\rho^2)}{4(\tau^2+\rho^2)^{3/2}}$, which runs from $+\frac{1}{2}$ at $\tau = -\infty$ to $-\frac{1}{2}$ at $\tau = +\infty$, hence carries topological charge $\nu = +1$. While this can also be interpreted as a localized pseudo-particle (fluctuation of gluonic field) in the Euclidean space, instanton acts as quantum tunneling from one QCD vacuum (with $N_{CS} = N$) to the neighbor one (with $N_{CS} = N + 1$) [26]. Anti-instantons are similar fluctuations but tunneling in the opposite direction.

While instantons correspond to quantum topological tunneling between QCD vacua, with the same energy between initial and final configuration, *sphaleron* is another topology tunneling solution [27] corresponding to "forced tunneling" that absorbs energy. In fact, the instantons and sphalerons not only change the topology structure of the gluon field, but also affect the Fermion field (quarks) via QCD vertices. In the following subsection we will discuss how fermion fields are influenced by instantons and what could be the possible effect.

1.2.3 Fermionic Axial Charge Fluctuation

Besides the continuous symmetries under Lorentz/gauge transformation, one is also interested in the discrete symmetric properties of a theory, under e.g. spatial reflection, time reversal, or charge conjugation. One quantity being interested in is the axial charge, which characterized the chirality imbalance and is a pseudo-scalar under space parity $\mathcal{P}$- and charge conjugation parity $\mathcal{CP}$- transformation. One can prove that such quantity conserves under classical QCD Euler–Lagrangian Equation (1.13). In other words, $\mathcal{P}$- and $\mathcal{CP}$-odd axial charge conserves at classical level. However, the renormalization of QCD (1.13) cannot be performed in a chirally invariant way [28, 29]. It means that the axial charge no longer conserves at quantum level, and such phenomenon is called *chiral anomaly* or *axial anomaly*.

As a result of the interplay of classical topologically nontrivial instanton solution and quantum chiral anomaly, the axial current, which includes N_f flavor degrees of freedom,

$$J_{5,\mu} \equiv \sum_f \overline{\psi}_f \gamma_\mu \gamma_5 \psi_f \tag{1.27}$$

is no longer conserved even in the massless limit ($m \to 0$):

$$\partial^\mu J_{5,\mu} = \left(\sum_f 2\mathrm{i}\, m_f \overline{\psi}_f \gamma_5 \psi_f \right) - \frac{g^2 N_f}{16\pi^2} G_a^{\mu\nu} \widetilde{G}_{a\mu\nu}. \tag{1.28}$$

Again, the second term is related to Chern–Simons current as shown in (1.24); hence in the massless limit ($m \to 0$), the change of the axial charge is associated with the topological charge

$$\Delta Q_5 = \int_{-\infty}^{\infty} \mathrm{d}t\, \frac{\partial}{\partial t} \int \mathrm{d}^3 x\, J_5^0$$

$$= -\frac{g^2 N_f}{16\pi^2} \int_{-\infty}^{\infty} \mathrm{d}t\, \frac{\mathrm{d}}{\mathrm{d}t} \int \mathrm{d}^3 x\, K^0$$

$$= 2N_f \left(N_{CS}|_{t=-\infty} - N_{CS}|_{t=+\infty} \right). \tag{1.29}$$

Consequently, being a $\mathcal{P}$- and $\mathcal{CP}$-odd quantity, the fermionic axial charge fluctuates according to the gluonic topology structure. In particular, it is worth noting that the g factor is the strong coupling coefficient, hence the axial charge fluctuates more when the coupling is stronger.

To sum it up, the $\mathcal{P}$- and $\mathcal{CP}$-odd axial charge is half-conserved in QCD process: its conservation is maintained at the classical level, but gets violated at quantum level. As a consequence of chiral anomaly and instantons with topologically nontrivial structure, one can expect locally fluctuating fermionic axial charge in QCD process, which serves as a smoking gun of both the local $\mathcal{P}$-, $\mathcal{CP}$-violation and the topologically nontrivial structure of classical QCD vacuum. On the other hand, being a half-conserved quantity, fermionic axial charge shall be conserved over global measurements,[2] and one can observe its violation only in event-wise

---

[2]Global measurement means measurement averaged over time, space and events.

observables. In Part I of this dissertation, we will discuss how we can observe such fluctuating axial charge by experiment signals caused by Chiral Magnetic Effect (CME) in relativistic heavy-ion collisions.

1.2.4 Solitons and Chromo-Magnetic-Monopole

In the above part of this section we discussed topologically nontrivial structures of pure Yang–Mills field, without any coupling to another (scalar/spinor) field. Now we extend our discussion to a gauge field coupled with self-interacting scalar fields

$$\mathcal{L} = -\frac{1}{4} F^a_{\mu\nu} F^{a\mu\nu} - \frac{1}{2} (D^\mu \varphi)^a (D_\mu \varphi)^a - \frac{\lambda}{8} (\varphi^a \varphi^a - v^2)^2. \tag{1.30}$$

For such theory, one can find the vacuum solution $\varphi^a = vn^a$, where n^a is a normalized "direction" vector in index space, satisfying $n^a n^a = 1$. With specified index direction n, the SU(N) symmetry is broken spontaneously. Noting that the theory is invariant under local gauge transformation, and one can always rotate the local index direction, with gauge field shifted correspondingly. In this case, we are interested in the case with non-vanishing winding number n_w, that gauge phase winded by n_w times if we wind a spatial round. However, such vacuum solution does not exist as there must be points where the field is ill-defined with unfixed gauge phase. Instead, one interesting question would be whether one could find a classical solution with finite energy that satisfies the boundary condition with non-vanishing winding number. An interesting solution with such characteristics is the *soliton* solution, in which the energy density is localized in space, does not dissipate or change in time. One example can be found in the simplified case, a SU(2) gauge field, i.e. focusing on the SU(2) sub-structure of a SU(3) gauge field. Also, we take the identity map that the gauge direction of the scalar field is "parallel" to spatial direction, hence the boundary condition of the scalar field is given by

$$\lim_{r \to \infty} \varphi^a(\boldsymbol{x}) = v x^a / r. \tag{1.31}$$

Such requirement leads to the solution as a *'t Hooft-Polyakov monopole* [30, 31]:

$$\varphi^a = v \left(\coth(evr) - \frac{1}{evr} \right) x^a / r \tag{1.32}$$

$$A^a_i = \left(1 - \frac{evr}{\sinh(evr)} \right) \varepsilon^{aij} x_j / (er^2). \tag{1.33}$$

From this solution one can further construct the chromo-magnetic field $B_i \equiv \frac{1}{2} \varepsilon_{ijk} F^a_{jk} \hat{\varphi}^a$, and then compute the magnetic flux through the boundary

$$\Phi = \int dS \cdot B = -\frac{4\pi}{e}, \tag{1.34}$$

With non-vanishing magnetic flux satisfying the Dirac charge quantization relation, one can see that the 't Hooft-Polyakov monopole is a solution containing chromo-magnetic-monopole.

Although this is not a classical solution of the *pure* Yang–Mills equation, the concept of chromo-magnetic-monopoles has attracted many theoretical interests [32–37]. Especially, Mandelstam and 't Hooft [33, 34] employed the chromo-magnetic-monopole (cmm) degree of freedom, as the quantum fluctuation of the gauge field, to explain the confinement of chromo-electric charges: in vacuum, the cmm degree of freedom is light and in the superconducting state, hence the chromo-electric degree of freedom is localized and constrained within flux tubes. Afterwards, further investigations [35, 38–42] are made to discuss how cmm's behave during QCD phase transition. It is believed that at the temperature much higher than phase transition temperature, $T \gg T_c$, the magnetic components have large effective mass and have negligible contribution to the dynamics of the system, while the electric components—quark and gluons—have small effective mass, and get liberated. When such a system cools down to the intermediate temperature $T \sim T_c$, the electric components become heavier and strongly coupled, while the magnetic components become more important to the evolution of the system. Given this case, one key question is raised naturally: could there be any measurable effects caused by such magnetic components?

The answer is positive. It is suggested [35, 40, 42, 43] that the chromo-magnetic-monopoles are possible to be observed via the medium effected on the initially produced hard partons, i.e., jet quenching observables. (Detailed description of jet quenching phenomena can be found in Sect. 1.3.3.) With different screening mass, the chromo-magnetic-monopoles induce different temperature dependence of the jet energy loss. Consequently, the chromo-magnetic-monopoles degree of freedom could influence the jet quenching observables, especially the azimuthal anisotropy. In Part II of this dissertation, we will quantitatively study how jet quenching observables could get influenced by the chromo-magnetic-monopole degree of freedom.

1.3   Soft and Hard Physics in Relativistic Heavy-Ion Collisions

As mentioned above, the best experimental environment to study a color-deconfined system so far is the quark–gluon plasma created in relativistic heavy-ion collisions, which have been intensively investigated at Brookhaven National Laboratory's Relativistic Heavy-Ion Collider (RHIC) [44, 45] as well as the CERN's Large Hadron Collider (LHC) [46, 47].

In the relativistic heavy-ion collision experiments, the participating nuclei are accelerated to extremely high energy, with Lorentz factor $\gamma \sim 10^2 - 10^3$. In such system, a huge amount of energy ($E \sim 10^2 - 10^3$ GeV) is deposited in tiny volume ($V \sim 10^2$ fm^3), and it creates a region with extremely high density of energy and particle number, and achieves the color-deconfinement phase according to lattice QCD calculation. It means that partons created in relativistic heavy-ion collisions are no longer confined in colorless hadrons, instead, they form a system that allows nonzero local color charges. In other words, in such experiments, it creates a new phase of matter consisting of color-deconfined quarks and gluons, and typically physicists called this new matter as quark–gluon plasma. On the other hand, given that in such system the mean free path of partons is smaller than the strong interaction scale, the medium created in the collisions is able to reach the state approaching local equilibrium. Consequently, one can treat such medium as a thermal system, and describe the collective behavior of the partons with thermal quantities.

The hot medium created in a relativistic heavy-ion collision experiences a rapid evolution. Soon after the collisions, a QCD Plasma droplet is created with high temperature ($T \sim 10^{12}$ K) and high pressure, while outside the droplet is the vacuum with zero pressure. The huge pressure gradient would push the medium to expand from its high-dense center to the low-dense edge, thus the hot medium cools down and gets diluted rapidly. Once the system gets cold and diluted enough, partons are confined again within colorless hadrons, and the system undergoes a transition from the deconfined plasma phase to the confined hadron phase. After the phase transition, the medium consisting of hadrons still expands and eventually becomes non-interacting hadrons. Finally, detectors at RHIC and LHC observe these particles, identify their species, and measure their momentum.

In practice, we learn the properties of the color-deconfined hot medium created in relativistic heavy-ion collisions by analyzing the behavior of particles produced, especially single particle distribution and multiple particle correlations. In particular, one common observable is the momentum differential information of particle production rate at different kinematic regimes of transverse momentum $p_T \equiv \sqrt{p_x^2 + p_y^2}$, azimuthal angle $\phi \equiv \arctan \frac{p_y}{p_x}$, and pseudo-rapidity $\eta \equiv \frac{1}{2} \ln \frac{|p|+p_z}{|p|-p_z}$, with the azimuthal distribution are conventionally expanded in series of Fourier harmonic components:

$$\frac{dN}{p_T dp_T d\phi d\eta} \equiv \frac{dN}{2\pi p_T dp_T d\eta} \left[1 + 2 \sum_{n=1} \right.$$

$$\left. \times \Big(a_n(\eta, p_T) \sin n(\phi - \Psi_n) + v_n(\eta, p_T) \cos n(\phi - \Psi_n) \Big) \right].$$

$$(1.35)$$

In Fig. 1.3 we show some examples of particle distribution observed in experiments, including: (left top) pseudo-rapidity dependence $dN_{ch}/d\eta$ from BRAHMS

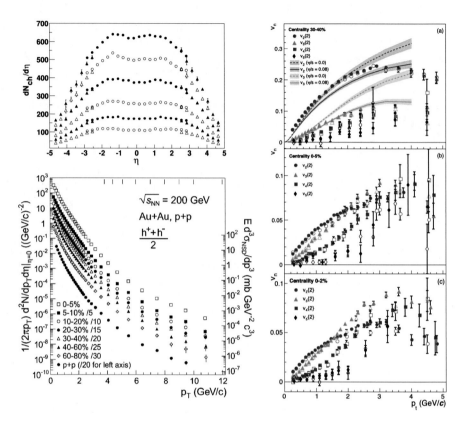

Fig. 1.3 Experimental results for charged particle distribution. (left top) From BRAHMS collaboration [48]: Distributions of $dN_{ch}/d\eta$ at $\sqrt{s_{NN}} = 200\,\text{GeV}$ Au–Au collisions, for centrality ranges of, top to bottom, 0–5%, 5–10%, 10–20%, 20–30%, 30–40%, and 40–50%. (left bottom) From STAR Collaboration [49]: Inclusive invariant p_T distributions of $(h^+ + h)/2$ for centrality-selected Au + Au and $p + p$ NSD interactions with beam energy $\sqrt{s_{NN}}$, $\sqrt{s} = 200$. (right) From ALICE collaboration [50]: v_2, v_3, v_4, v_5 as a function of transverse momentum and for three event centralities in $\sqrt{s_{NN}} = 2.76$ TeV Pb–Pb collisions. The full, open symbols are for $\Delta\eta > 0.2$ and $\Delta\eta > 1.0$, respectively. (a) 30–40% compared to hydrodynamic model calculations, (b) 0–5% centrality percentile, and (c) 0–2% centrality percentile

collaboration [48], (left bottom) invariant p_T distributions from STAR Collaboration [49], (right) and flow harmonics v_n from ALICE collaboration [50]. One can see that the pseudo-rapidity distribution is quite flat within the range of $-2 \lesssim \eta \lesssim 2$. Such feature can be understood in the way that the huge pressure gradient along longitudinal direction pushes the plasma to rapidly expand along this direction, with fluid velocity $v_z \sim z/t$. Such behavior can be translated to be $v_\eta \sim 0$ in the propertime-rapidity frame $\tau \equiv \sqrt{t^2 - z^2}$, $\eta \equiv \frac{1}{2} \ln \frac{t+z}{t-z}$, and leads to the boost-invariance of the system. Hence, typically in experiments, one can focus on the middle rapidity range, e.g., $|\eta| < 0.5$, and measure the p_T dependence of spectra $\frac{dN}{2\pi p_T dp_T}$ as well as flow harmonics v_n around $\eta = 0$.

On the other hand, it is worth noting that the local temperature of the medium varies around 100–600 MeV, with the phase transition temperature $T_c \approx 165$ MeV, thus, most of the particles are produced with transverse momentum $p_T < 3$ GeV. Typically final detected hadrons with $p_T \lesssim 3$ GeV are called *soft particles*, which are the products of collective movement of the plasma. On the other hand, rare final hadrons with $p_T \gtrsim 10$ GeV, typically referred to as *hard particles*, are produced predominately by initial hard scattering. In the following subsections we will discuss how one can understand the behavior of these soft and hard particles with sophisticated phenomenological models.

1.3.1 Nearly-Perfect Fluidity

One fascinating property of the hot medium created in relativistic heavy-ion collisions is that it evolves as nearly-perfect fluid, which means one can model the medium as relativistically expanding fluid with small viscosity, and explain experimental results of particle distribution at lower transverse momentum regime ($p_T \lesssim 3$ GeV), including p_T spectrum and varying orders of azimuthal harmonics $v_n(p_T)$. This property has been validated by many independent tests from different groups by using their own simulation packages, e.g., VISHNU [51–54], MUSIC [55, 56], CLVisc [57, 58], vUSP [59, 60], ECHO-QGP [61–63], etc.

In numerical simulations of relativistic hydrodynamics, it adopts the Israel–Stewart framework [64] for the second-order viscous hydrodynamic equations, which describes the evolution of the hot medium based on energy-momentum conservation and charge conservation:

$$\hat{D}_\nu T^{\mu\nu} = 0, \tag{1.36}$$

$$\hat{D}_\mu J_i^\mu = 0, \tag{1.37}$$

with the fluid energy-momentum tensor $T^{\mu\nu}$ and charge currents J_i^μ given by:

$$T^{\mu\nu} = \varepsilon\, u^\mu u^\nu - (p + \Pi)\, \Delta^{\mu\nu} + \pi^{\mu\nu}, \tag{1.38}$$

$$J_i^\mu = n_i u^\mu + v_i^\mu, \tag{1.39}$$

where thermal quantities like energy density ε, pressure p and conserved charge densities n_i are related via the equation of state, while bulk pressure Π, shear stress tensor $\pi^{\mu\nu}$, and diffusion currents v_i^μ take the non-equilibrium effects into account. They relax towards the corresponding Navier–Stokes form with shear viscosity η, bulk viscosity ζ, and relaxation times τ_π, τ_Π, τ_v,

$$\Delta^{\mu\alpha}\Delta^{\nu\beta}\hat{d}\pi_{\alpha\beta} = -\frac{1}{\tau_\pi}(\pi^{\mu\nu} - 2\eta\sigma^{\mu\nu}) - \frac{\pi^{\mu\nu}}{2}\frac{\eta T}{\tau_\pi}\hat{D}_\lambda\left(\frac{\tau_\pi}{\eta T}u^\lambda\right), \tag{1.40}$$

$$\hat{d}\Pi = -\frac{1}{\tau_\Pi}(\Pi + \zeta\theta) - \frac{\Pi}{2}\frac{\zeta T}{\tau_\Pi}\hat{D}_\lambda\left(\frac{\tau_\Pi}{\zeta T}u^\lambda\right), \qquad (1.41)$$

$$\Delta^\mu_\nu\hat{d}\left(v^\nu_i\right) = -\frac{1}{\tau_\nu}\left[\left(v^\mu_i\right) - \frac{\sigma}{2}T\Delta^{\mu\nu}\partial_\nu\left(\frac{\mu_i}{T}\right)\right]. \qquad (1.42)$$

After specifying suitable initial conditions, one can then obtain the space-time profile of all the thermal quantities, as well as the fluid velocity of the medium by solving the above differential equations. At the end of the fluid evolution, the QGP hadronizes at a specific temperature—the freeze-out temperature T_f—and the final hadrons are then locally produced in all fluid cells on the freeze-out hyper-surface with a local thermal distribution including non-equilibrium corrections (see e.g. [54] for details), following the Cooper-Frye freeze-out formula [65]

$$E\frac{dN}{d^3p}(x^\mu, p^\mu) = \frac{g}{(2\pi)^3}\int_{\Sigma_{fo}}p^\mu d^3\sigma_\mu f(x, p). \qquad (1.43)$$

Numerical hydrodynamic simulations indicate that experimental results of particle distribution can be well explained by taking the shear viscosity to entropy ratio η/s with value of 0.1–0.2. Such value approaches the lower limit that $(\eta/s)_{min} = 1/4\pi$ determined by AdS-CFT calculation [66], and this indicates that hot medium created in relativistic heavy-ion collisions is likely to be the most perfect fluid we have ever observed. By treating the hot medium as nearly-perfect fluid, relativistic hydrodynamic simulations provide good understanding of collective behaviors of low p_T particles, namely "bulk evolution."

It might be worth noting that such perfectness of the fluid reflects strong coupling between particles in this temperature range. Given the strong interactions between particles, the average-free-path of each particle is too short and particle exchange between fluid shears gets reduced, and consequently the shear viscosity is small. As mentioned above, another important feature in the strongly coupling system is the topologically nontrivial objects, especially topological charge which could be reflected in fermion axial charge fluctuation, as well as the chromo-magnetic-monopole degree of freedom. In stronger coupling case, the fermion axial charge fluctuates more, while the chromo-magnetic-monopoles play more important role in the opacity of the fluid. Based on this data validated description of the bulk background, one can further study anomalous transport as well as jet quenching phenomena as the probes of QCD nontrivial topological structures.

1.3.2   Probing the Fluctuating Axial Charges in Heavy-Ion Collisions

As mentioned above, a main goal of this dissertation is to study the effects of fluctuating fermionic axial charges, which is the interplay of chiral anomaly and

QCD topological charges. However, particle detectors at RHIC and LHC are blind to pseudo-scalar/axial vector observables, e.g., particle spin, and the most promising observables come from the effect of vector currents, as the product of aforementioned axial charges and another axial vector field. Fortunately, in heavy-ion collisions we do have global axial vector fields.

Magnetic Field One example of axial vector field one could expect in heavy-ion collisions is the magnetic field. To estimate its strength, let us consider a point particle with electric charge q moving with velocity $v\hat{z}$, which generates a magnetic field

$$\mathbf{B}(t, x, y, z) = (q\, v\, \gamma) \frac{(y - y_0)\hat{x} - (x - x_0)\hat{y}}{[(x - x_0)^2 + (y - y_0)^2 + \gamma^2(z - z_0 - vt)^2]^{3/2}}, \quad (1.44)$$

where (x_0, y_0, z_0) denotes the position of q at $t = 0$. Benefitted from the large Lorentz γ-factor, as well as the fact that B_y has the same sign in the overlap region of two opposite-moving nuclei, one could expect an ultra-strong magnetic field $B_y(t = 0) \sim 5m_\pi^2/e \sim 10^{15}$ T for middle-central Au–Au collisions at $\sqrt{s_{NN}} = 200$ GeV. Such strong magnetic field allows us to study many anomalous effects, e.g., Chiral Magnetic Effect [67–70], Chiral Magnetic Wave [71–73]. In the meanwhile, some other effects like modification on charged particle elliptic flow [74–77] are predicted to serve as independent tests of the magnetic field. It is worth mentioning that spectators—particles which are not deposited in the medium but carries most of the energies and electric charges—leave with nearly speed of light after the collision, hence the external magnetic field decays rapidly (with live time $\tau \sim R_{\text{nuclei}}/\gamma \sim 0.1$ fm/c). On the other hand, the quark–gluon plasma is a conducting plasma, and therefore generates induction current in response to the fast changing magnetic field. In principle such mechanism could delay the decrease of the magnetic field. While many efforts have been made to compute the time dependence of the magnetic field [78–81], the answers from different studies vary considerably. To fully address this issue, one needs to treat both the medium and the magnetic field as dynamically evolving together, and solve the full magneto-hydrodynamics equations.

Subatomic Swirls Another example of axial vector is the angular momentum, which relates to the rotation of a system. In fact, the system created in heavy-ion collisions is fast-rotating. In a typical non-central collision, the two opposite-moving nuclei have their center-of-mass misaligned and thus carry a considerable angular momentum $L = b\, p \sim 10^{4\sim6}\, \hbar$. It has been long believed that a good fraction of this angular momentum will remain in the created hot plasma and lead to strong nonzero vortical fluid structures (often quantified by fluid vorticity) during its hydrodynamic evolution, thus it forms the "subatomic swirls." A possible signature of the fluid vorticity is the spin polarization of the produced particles which on average should be aligned with the colliding system's global angular momentum direction. Recently the STAR Collaboration reported their remarkable discovery of the "subatomic swirls" in the high-energy Au–Au collisions [82]. By a clever analysis of spin orientation of the Λ hyperons, the STAR Collaboration was able to find very strong evidence for the global polarization effect, from which they extracted an average fluid vorticity of about 10^{21} s^{-1}, being the most vortical fluid ever known.

With these strong axial vector fields—magnetic field and vorticity field—one would be able to observe the fluctuating fermionic axial charge via vector current. In this dissertation we will focus on quantitative study of the Chiral Magnetic Effect (for detailed descriptions of this effect, see Chap. 2), by employing our newly developed Anomalous-Viscous Fluid Dynamics (AVFD) simulation framework. In Chap. 7 we will also discuss about effects of rotation.

1.3.3   Jets—Tomography of the QCD Plasma

With the low-p_T particle distributions well explained by relativistic hydrodynamics modeling, let us continue to discuss about hard particles with high transverse momentum $p_T \gtrsim 10$ GeV, which is much higher than the thermal energy scale of the plasma. The hard particles are produced dominantly by fragmentation from initially generated hard partons, rather than re-combination of partons in the hot Plasma. In experiments, the final hadrons fragmented from a high p_T parton are typically distributed in a narrow cone, and such a narrow cone of hadrons are referred to as a *jet*. Typically we extend this definition, and a "jet" also refers to the high p_T parton itself.

As mentioned above, the jet physics is in the large energy scale corresponding to small coupling constant, and can be computed by perturbative QCD calculation. Hence, in $p + p$ and $e^+ + e^-$ collisions, the high p_T observables serve as a crucial test of our knowledge of strong interaction. Nevertheless, in nucleus–nucleus collisions, although hard hadrons come from jets, high-p_T observables are not irrelevant with the hot medium. In fact, an initially generated hard parton passes through the expanding hot medium before it finally fragments into colorless hadrons. In the medium, the hard parton interacts with soft particles and loses energy due to its radiating gluons and elastically scattering with the soft particles. Benefitted from its large energy scale, where the strong interaction reaches the small coupling constant regime, a jet has a good chance to survive and eventually reaches the boundary of the plasma, rather than being completely stopped and absorbed by the medium. The phenomenon that the energy of a jet get reduced due to its interaction with the hot medium is called *jet quenching*. Jet quenching observables serve as the tomography of the QCD plasma and provide independent test of our knowledge to the hot medium created in relativistic heavy-ion collisions.

Regarding to initial hard processes, a nucleus–nucleus (AA) collision can be treated as the combination of multiple individual nucleon–nucleon (pp) collisions. One can expect that the p_T-depending production rate of initial hard partons, before quenching or fragmentation, in AA collisions should be proportional to that in pp collisions, with proportionality constant being the number of binary nucleon–nucleon collisions N_{bin}. Consequently, one can characterize the distribution of unquenched/initial jet observables as N_{bin} times of that in pp collisions, and observe the nuclear modification factor as the ratio of (p_T, ϕ)-differential hadron production rate for quenched jets to that for unquenched jets:

$$R_{AA}(p_T, \phi) \equiv \frac{dN_{AA}/(dp_T d\phi)}{N_{bin} \cdot dN_{pp}/(dp_T d\phi)}. \tag{1.45}$$

Again, typically the azimuthal dependence is usually characterized by Fourier harmonics, with Ψ_2 (Ψ_3) being the second (third)-order event plane determined by soft particle distribution:

$$R_{AA}(p_T, \phi) \equiv R_{AA}(p_T) \times \left[1 + 2 v_2(p_T) \cos 2(\phi - \Psi_2) \right.$$

$$\left. + 2 v_3(p_T) \cos 3(\phi - \Psi_3) + \cdots \right]. \tag{1.46}$$

In AA collisions, a final state jet is the quenching result of an initial parton with higher energy, hence it has lower production rate than unquenched jets with same p_T. Consequently, R_{AA} is smaller than unity (in high transverse momentum regime $p_T \gtrsim 10\,\text{GeV}$). The suppression of R_{AA} characterize the opacity/transparency of the medium to jets. On the other hand, azimuthal harmonics v_2 and v_3 come from the difference in path length and medium density along different directions, and characterize the anisotropy of the hot medium. We expect the jet quenching observables R_{AA}, v_2, v_3 could provide more information about the constituent of the QCD plasma created in relativistic heavy-ion collisions, and serve as a hard probe of the aforementioned chromo-magnetic-monopole charges in the plasma.

To summarize, in relativistic heavy-ion collision experiments it creates a new phase of matter, consisting of color-deconfined quarks and gluons, and provides a special environment to study strong interaction, especially color confinement. Besides, in such collisions, it produces the hottest matter ($T \sim 10^{12}\,\text{K}$), the strongest magnetic field ($B \sim 10^{16}\,\text{T}$), the most vortical system ($\omega \sim 10^{21}\,\text{Hz}$), and likely the most perfect fluid ($\eta/s \sim 0.1$) in the observed universe. In this dissertation, we aim to take advantage of such extreme environment, to study the phenomena caused by topologically nontrivial structures of QCD, especially the Chiral Magnetic Effect and jet quenching in a QCD plasma with chromo-magnetic-monopoles.

1.4 Dissertation Organization

In this dissertation we focus on both soft and hard probes of color-deconfined medium, especially its topological properties, in relativistic heavy-ion collisions. This thesis summarizes the results published in Refs. [83–89], along with some on-going researches. In addition to the introduction here in Chap. 1 and the summary in Chap. 11, the main content of this dissertation could be divided into two parts, focusing on soft and hard observables, respectively.

In Part I, we focus on quantitative study of the Chiral Magnetic Effect, as the soft probe of topological charge transition. In Chap. 2 a brief discussion on the relevant experimental observables, which are also the quantities to be computed, will be given. The detailed implementation of the Anomalous-Viscous Fluid Dynamics (AVFD) framework will be described in Chap. 3. Then Chap. 4 presents the quantitative results for the CME signals on top of event-averaged hydro background, while Chap. 5 will discuss how CME signals are influenced by event-by-event fluctuations. Finally we extend the study to consider chiral effects in pre-equilibrium stage in Chap. 6, and discuss the rotation of the QCD fluid, as well as vortical effects in Chap. 7.

In Part II we focus on jet quenching phenomena, the hard probe of constituent of the QCD fluid. In Chap. 8, we briefly introduce how we simulate, by using the CUJET package, the jet energy loss in the relativistically expanding QCD fluid. Then in Chap. 9 we present a systematic study of the jet quenching observables on top of an event-averaged background. Finally in Chap. 10 we discuss how jet observables are influenced by the fluctuating bulk background.

References

1. J.S. Schwinger, Phys. Rev. **74**, 1439 (1948). https://doi.org/10.1103/PhysRev.74.1439
2. J.S. Schwinger, Phys. Rev. **73**, 416 (1948). https://doi.org/10.1103/PhysRev.73.416
3. R.P. Feynman, Phys. Rev. **76**, 769 (1949). https://doi.org/10.1103/PhysRev.76.769
4. R.P. Feynman, Phys. Rev. **80**, 440 (1950). https://doi.org/10.1103/PhysRev.80.440
5. H.A. Bethe, Phys. Rev. **72**, 339 (1947). https://doi.org/10.1103/PhysRev.72.339
6. C.N. Yang, R.L. Mills, Phys. Rev. **96**, 191 (1954). https://doi.org/10.1103/PhysRev.96.191
7. M. Gell-Mann, Phys. Rev. **125**, 1067 (1962). https://doi.org/10.1103/PhysRev.125.1067
8. D.J. Gross, F. Wilczek, Phys. Rev. **D8**, 3633 (1973). https://doi.org/10.1103/PhysRevD.8.3633
9. H.D. Politzer, Phys. Rev. Lett. **30**, 1346 (1973). https://doi.org/10.1103/PhysRevLett.30.1346
10. C. Patrignani et al., Chin. Phys. **C40**(10), 100001 (2016). https://doi.org/10.1088/1674-1137/40/10/100001
11. M.M. Aggarwal et al. (2010). arXiv:1007.2613
12. B.I. Abelev et al., Phys. Rev. **C75**, 054906 (2007). https://doi.org/10.1103/PhysRevC.75.054906
13. J. Adams et al., Phys. Rev. Lett. **95**, 152301 (2005). https://doi.org/10.1103/PhysRevLett.95.152301
14. C. Baglin et al., Phys. Lett. **B220**, 471 (1989). https://doi.org/10.1016/0370-2693(89)90905-2
15. C. Baglin et al., Phys. Lett. **B255**, 459 (1991). https://doi.org/10.1016/0370-2693(91)90795-R
16. S. Gupta, X. Luo, B. Mohanty, H.G. Ritter, N. Xu, Science **332**, 1525 (2011). https://doi.org/10.1126/science.1204621
17. L. Adamczyk et al., Phys. Rev. Lett. **112**, 032302 (2014). https://doi.org/10.1103/PhysRevLett.112.032302

18. L. Adamczyk et al., Phys. Rev. Lett. **113**, 092301 (2014). https://doi.org/10.1103/PhysRevLett.113.092301
19. X. Luo, N. Xu, Nucl. Sci. Tech. **28**(8), 112 (2017). https://doi.org/10.1007/s41365-017-0257-0
20. M.A. Stephanov, Phys. Rev. Lett. **107**, 052301 (2011). https://doi.org/10.1103/PhysRevLett.107.052301
21. S.S. Chern, J. Simons, Annals Math. **99**, 48 (1974). https://doi.org/10.2307/1971013
22. E. Witten, Commun. Math. Phys. **117**, 353 (1988). https://doi.org/10.1007/BF01223371
23. A.A. Belavin, A.M. Polyakov, A.S. Schwartz, Yu.S. Tyupkin, Phys. Lett. **B59**, 85 (1975). https://doi.org/10.1016/0370-2693(75)90163-X
24. G. 't Hooft, Phys. Rev. Lett. **37**, 8 (1976). https://doi.org/10.1103/PhysRevLett.37.8
25. G. 't Hooft, Phys. Rev. **D14**, 3432 (1976). https://doi.org/10.1103/PhysRevD.18.2199.3, https://doi.org/10.1103/PhysRevD.14.3432
26. A.M. Polyakov, Nucl. Phys. **B120**, 429 (1977). https://doi.org/10.1016/0550-3213(77)90086-4
27. D.M. Ostrovsky, G.W. Carter, E.V. Shuryak, Phys. Rev. **D66**, 036004 (2002). https://doi.org/10.1103/PhysRevD.66.036004
28. S.L. Adler, Phys. Rev. **177**, 2426 (1969). https://doi.org/10.1103/PhysRev.177.2426
29. J.S. Bell, R. Jackiw, Nuovo Cim. **A60**, 47 (1969). https://doi.org/10.1007/BF02823296
30. G. 't Hooft, Nucl. Phys. **B79**, 276 (1974). https://doi.org/10.1016/0550-3213(74)90486-6
31. A.M. Polyakov, JETP Lett. **20**, 194 (1974)
32. C. Montonen, D.I. Olive, Phys. Lett. **72B**, 117 (1977). https://doi.org/10.1016/0370-2693(77)90076-4
33. S. Mandelstam, Phys. Rept. **23**, 245 (1976). https://doi.org/10.1016/0370-1573(76)90043-0
34. G. 't Hooft, Nucl. Phys. **B190**, 455 (1981). https://doi.org/10.1016/0550-3213(81)90442-9
35. J. Liao, E. Shuryak, Phys. Rev. **C75**, 054907 (2007). https://doi.org/10.1103/PhysRevC.75.054907
36. M.N. Chernodub, V.I. Zakharov, Phys. Rev. Lett. **98**, 082002 (2007). https://doi.org/10.1103/PhysRevLett.98.082002
37. A. D'Alessandro, M. D'Elia, Nucl. Phys. **B799**, 241 (2008). https://doi.org/10.1016/j.nuclphysb.2008.03.002
38. J. Liao, E. Shuryak, Phys. Rev. **C77**, 064905 (2008). https://doi.org/10.1103/PhysRevC.77.064905
39. J. Liao, E. Shuryak, Phys. Rev. **D82**, 094007 (2010). https://doi.org/10.1103/PhysRevD.82.094007
40. J. Liao, E. Shuryak, Phys. Rev. Lett. **101**, 162302 (2008). https://doi.org/10.1103/PhysRevLett.101.162302
41. J. Liao, E. Shuryak, Phys. Rev. Lett. **102**, 202302 (2009). https://doi.org/10.1103/PhysRevLett.102.202302
42. J. Liao, E. Shuryak, Phys. Rev. Lett. **109**, 152001 (2012). https://doi.org/10.1103/PhysRevLett.109.152001
43. B.G. Zakharov, JETP Lett. **88**, 781 (2008). https://doi.org/10.1134/S0021364008240016
44. M. Harrison, S.G. Peggs, T. Roser, Ann. Rev. Nucl. Part. Sci. **52**, 425 (2002). https://doi.org/10.1146/annurev.nucl.52.050102.090650
45. M. Harrison, T. Ludlam, S. Ozaki, Nucl. Instrum. Meth. **A499**, 235 (2003). https://doi.org/10.1016/S0168-9002(02)01937-X
46. CERN, CERN-Yellow-91-03 (1991)
47. S. Myers, Int. J. Mod. Phys. **A28**, 1330035 (2013). https://doi.org/10.1142/S0217751X13300354
48. I.G. Bearden et al., Phys. Rev. Lett. **88**, 202301 (2002). https://doi.org/10.1103/PhysRevLett.88.202301
49. J. Adams et al., Phys. Rev. Lett. **91**, 172302 (2003). https://doi.org/10.1103/PhysRevLett.91.172302

50. K. Aamodt et al., Phys. Rev. Lett. **107**, 032301 (2011). https://doi.org/10.1103/PhysRevLett. 107.032301
51. H. Song, U.W. Heinz, Phys. Rev. **C77**, 064901 (2008). https://doi.org/10.1103/PhysRevC.77. 064901
52. H. Song, U.W. Heinz, Phys. Rev. **C78**, 024902 (2008). https://doi.org/10.1103/PhysRevC.78. 024902
53. H. Song, S. Bass, U.W. Heinz, Phys. Rev. **C89**(3), 034919 (2014). https://doi.org/10.1103/ PhysRevC.89.034919
54. C. Shen, Z. Qiu, H. Song, J. Bernhard, S. Bass, U. Heinz, Comput. Phys. Commun. **199**, 61 (2016). https://doi.org/10.1016/j.cpc.2015.08.039
55. B. Schenke, S. Jeon, C. Gale, Phys. Rev. Lett. **106**, 042301 (2011). https://doi.org/10.1103/ PhysRevLett.106.042301
56. C. Gale, S. Jeon, B. Schenke, P. Tribedy, R. Venugopalan, Phys. Rev. Lett. **110**(1), 012302 (2013). https://doi.org/10.1103/PhysRevLett.110.012302
57. L. Pang, Q. Wang, X.N. Wang, Phys. Rev. **C86**, 024911 (2012). https://doi.org/10.1103/ PhysRevC.86.024911
58. L.G. Pang, H. Petersen, X.N. Wang, Phys. Rev. **C97**(6), 064918 (2018). https://doi.org/10. 1103/PhysRevC.97.064918
59. J. Noronha-Hostler, G.S. Denicol, J. Noronha, R.P.G. Andrade, F. Grassi, Phys. Rev. **C88**(4), 044916 (2013). https://doi.org/10.1103/PhysRevC.88.044916
60. J. Noronha-Hostler, J. Noronha, F. Grassi, Phys. Rev. **C90**(3), 034907 (2014). https://doi.org/ 10.1103/PhysRevC.90.034907
61. L. Del Zanna, V. Chandra, G. Inghirami, V. Rolando, A. Beraudo, A. De Pace, G. Pagliara, A. Drago, F. Becattini, Eur. Phys. J. **C73**, 2524 (2013). https://doi.org/10.1140/epjc/s10052-013-2524-5
62. S. Floerchinger, U.A. Wiedemann, A. Beraudo, L. Del Zanna, G. Inghirami, V. Rolando, Phys. Lett. **B735**, 305 (2014). https://doi.org/10.1016/j.physletb.2014.06.049
63. F. Becattini, G. Inghirami, V. Rolando, A. Beraudo, L. Del Zanna, A. De Pace, M. Nardi, G. Pagliara, V. Chandra, Eur. Phys. J. **C75**(9), 406 (2015). https://doi.org/10.1140/epjc/ s10052-015-3624-1
64. W. Israel, J.M. Stewart, Annals Phys. **118**, 341 (1979). https://doi.org/10.1016/0003-4916(79)90130-1
65. F. Cooper, G. Frye, Phys. Rev. **D10**, 186 (1974). https://doi.org/10.1103/PhysRevD.10.186
66. P. Kovtun, D.T. Son, A.O. Starinets, Phys. Rev. Lett. **94**, 111601 (2005). https://doi.org/10. 1103/PhysRevLett.94.111601
67. A. Vilenkin, Phys. Rev. **D22**, 3080 (1980). https://doi.org/10.1103/PhysRevD.22.3080
68. M.A. Metlitski, A.R. Zhitnitsky, Phys. Rev. **D72**, 045011 (2005). https://doi.org/10.1103/ PhysRevD.72.045011
69. G.M. Newman, D.T. Son, Phys. Rev. **D73**, 045006 (2006). https://doi.org/10.1103/PhysRevD. 73.045006
70. D.E. Kharzeev, L.D. McLerran, H.J. Warringa, Nucl. Phys. **A803**, 227 (2008). https://doi.org/ 10.1016/j.nuclphysa.2008.02.298
71. D.E. Kharzeev, H.U. Yee, Phys. Rev. **D83**, 085007 (2011). https://doi.org/10.1103/PhysRevD. 83.085007
72. E.V. Gorbar, V.A. Miransky, I.A. Shovkovy, Phys. Rev. **D83**, 085003 (2011). https://doi.org/ 10.1103/PhysRevD.83.085003
73. Y. Burnier, D.E. Kharzeev, J. Liao, H.U. Yee, Phys. Rev. Lett. **107**, 052303 (2011). https:// doi.org/10.1103/PhysRevLett.107.052303
74. K. Tuchin, J. Phys. **G39**, 025010 (2012). https://doi.org/10.1088/0954-3899/39/2/025010
75. G. Basar, D. Kharzeev, D. Kharzeev, V. Skokov, Phys. Rev. Lett. **109**, 202303 (2012). https:// doi.org/10.1103/PhysRevLett.109.202303
76. G. Basar, D.E. Kharzeev, E.V. Shuryak, Phys. Rev. **C90**(1), 014905 (2014). https://doi.org/ 10.1103/PhysRevC.90.014905

77. X. Guo, S. Shi, N. Xu, Z. Xu, P. Zhuang, Phys. Lett. **B751**, 215 (2015). https://doi.org/10.1016/j.physletb.2015.10.038
78. L. McLerran, V. Skokov, Nucl. Phys. **A929**, 184 (2014). https://doi.org/10.1016/j.nuclphysa.2014.05.008
79. U. Gursoy, D. Kharzeev, K. Rajagopal, Phys. Rev. **C89**(5), 054905 (2014). https://doi.org/10.1103/PhysRevC.89.054905
80. K. Tuchin, Phys. Rev. **C93**(1), 014905 (2016). https://doi.org/10.1103/PhysRevC.93.014905
81. G. Inghirami, L. Del Zanna, A. Beraudo, M.H. Moghaddam, F. Becattini, M. Bleicher, Eur. Phys. J. **C76**(12), 659 (2016). https://doi.org/10.1140/epjc/s10052-016-4516-8
82. L. Adamczyk et al., Nature **548**, 62 (2017). https://doi.org/10.1038/nature23004
83. Y. Jiang, S. Shi, Y. Yin, J. Liao, Chin. Phys. **C42**(1), 011001 (2018). https://doi.org/10.1088/1674-1137/42/1/011001
84. S. Shi, Y. Jiang, E. Lilleskov, J. Liao, Annals Phys. **394**, 50 (2018). https://doi.org/10.1016/j.aop.2018.04.026
85. A. Huang, S. Shi, Y. Jiang, J. Liao, P. Zhuang, Phys. Rev. **D98**(3), 036010 (2018). https://doi.org/10.1103/PhysRevD.98.036010
86. A. Huang, Y. Jiang, S. Shi, J. Liao, P. Zhuang, Phys. Lett. **B777**, 177 (2018). https://doi.org/10.1016/j.physletb.2017.12.025
87. S. Shi, K. Li, J. Liao, Phys. Lett. **B788**, 409 (2019). https://doi.org/10.1016/j.physletb.2018.09.066
88. S. Shi, J. Liao, M. Gyulassy, Chin. Phys. **C42**(10), 104104 (2018). https://doi.org/10.1088/1674-1137/42/10/104104
89. S. Shi, J. Liao, M. Gyulassy, Chin. Phys. **C43**(4), 044101 (2019). https://doi.org/10.1088/1674-1137/43/4/044101

Part I
Soft Probe of Topological Charge Transition

Chapter 2
The Chiral Magnetic Effect and Corresponding Observables in Heavy-Ion Collisions

2.1 The Chiral Magnet Effect

Symmetry principles play instrumental roles in the construction of our most basic physical theories. A special category of "symmetry" is the so-called anomaly, which is a well-defined classical symmetry of a theory but gets broken at quantum level. A most famous example of anomaly is the chiral anomaly which is a very fundamental aspect of quantum theories with spin-$\frac{1}{2}$ chiral fermion, from the Standard Model to supersymmetric field theories or even string theories. In such theories, the classical conservation law for the right-handed (RH) or left-handed (LH) chiral current $J_{R/L}^{\mu}$ gets broken at quantum level when coupled to (Abelian or non-Abelian) gauge fields. The famous Adler–Bell–Jackiw anomaly for one species of chiral fermion (with electric charge Q_f) in electromagnetic fields is given by Adler [1], Bell, and Jackiw [2]

$$\partial_\mu J_{R/L}^{\mu} = \pm C_A \mathbf{E} \cdot \mathbf{B}, \tag{2.1}$$

where $C_A = \frac{Q_f^2}{4\pi^2}$ is a universal coefficient. In the case of non-Abelian anomaly, one simply replaces the electromagnetic fields by appropriate non-Abelian gauge fields with slight modification of the constant C_A.

Microscopic symmetry principles also manifest themselves nontrivially in macroscopic physics. For example, the fluid dynamics as a general long-time large-distance effective description of any macroscopic system is a direct manifestation of symmetries in the underlying microscopic dynamics. The fluid dynamical equations for energy-momentum tensor and for charged currents are the direct consequences of conservation of energy, momentum, and charge which all originate from corresponding microscopic symmetries.

© Springer Nature Switzerland AG 2019
S. Shi, *Soft and Hard Probes of QCD Topological Structures in Relativistic Heavy-Ion Collisions*, Springer Theses,
https://doi.org/10.1007/978-3-030-25482-7_2

This naturally brings up a deep question: what are the implications of microscopic quantum anomaly (as a sort of "half symmetry") on the macroscopic properties of matter? Such question has triggered significant interest and important progress recently. As it turns out, unusual macroscopic transport currents can be induced by chiral anomaly under suitable conditions, with the notable example of the Chiral Magnetic Effect (CME) where a vector current (e.g., electric current $\mathbf{J}$) is generated along an external magnetic field $\mathbf{B}$ [3–7]:

$$\mathbf{J} = C_A \mu_5 \mathbf{B}, \tag{2.2}$$

where the quantity μ_5 is a chiral chemical potential that quantifies the macroscopic imbalance between RH and LH fermions in the system. Most remarkably, the conducting coefficient of this current C_A, being as the anomaly coefficient C_A in Eq. (2.1), is totally dictated by the microscopic anomaly relation. The other highly nontrivial feature of the CME is that the anomalous transport process underlying this quantum current in the above equation is time-reversal even, i.e., non-dissipative [8].

In the context of strong interaction physics as described by Quantum Chromo-Dynamics (QCD), the chiral anomaly provides a unique access to the topological configurations such as instantons and sphalerons which are known to play crucial roles in many non-perturbative phenomena [9, 10]. In particular, accompanying such topological configurations are the fluctuations of chirality imbalance (i.e., difference in the number of RH versus LH quarks) in the system, precisely due to the anomaly relation. That is, the macroscopic chirality fluctuations of fermions in the system, which would be experimentally measurable, reflect directly the gluonic topological fluctuations, the information about which would be otherwise unaccessible. Furthermore, the CME (2.2) provides a nontrivial way to manifest a nonzero macroscopic chirality (as quantified by μ_5) in QCD matter: a nonzero μ_5 with the presence of an external magnetic field $\mathbf{B}$ would be measurable via the induced electric current $\mathbf{J}$.

The experimental realization of the CME has been enthusiastically pursued in two very different types of real world materials. The first is the Dirac and Weyl semimetals in condensed matter physics [11–16] where the CME has been successfully observed. The other is the quark–gluon plasma (QGP) in heavy-ion collision experiments at the Relativistic Heavy Ion Collider (RHIC) and the Large Hadron Collider (LHC) [17–23]: see recent reviews in, e.g., [24–27]. Encouraging evidence of CME-induced charge separation signals in those collisions have been reported, albeit with ambiguity due to background contamination. Crucial for addressing such issue is the need of quantitative predictions for CME signals with sophisticated modelings. As a crucial step toward achieving this goal, we develop the Anomalous-Viscous Fluid Dynamics (AVFD) framework [28], which implements the fluid dynamical evolution of chiral fermion currents (of the light flavor quarks) with anomalous transport in QGP on top of the expanding (neutral) bulk background described by the VISH2+1 hydrodynamic simulations. With this newly developed tool, we report our systematic and quantitative investigations on the CME signals in heavy-ion collisions as well as the influence of various existing theoretical uncertainties.

2.2 Charge Separation Measurements in Heavy-Ion Collisions

In relativistic heavy-ion collisions the nuclei are accelerated to travel at nearly the speed of light. The strong Lorentz boost effect changes the nuclear geometry as observed in the laboratory frame, and more importantly, leads to a very strong magnetic field, which is Lorentz-transformed from the Coulomb-like electric field in the nucleus' co-moving frame. In a typical off-central collisions, a magnetic field along the out-of-plane direction (conventionally defined as the $\hat{y}$-direction) is in the overlap region, where the hot quark–gluon plasma (QGP) also forms. For a given collision event with chirality imbalance (e.g., say, with more RH than LH light quarks corresponding to a positive chiral chemical potential $\mu_5 > 0$), the CME-induced electric current $\mathbf{J}$ as in Eq. (2.1) is expected and it transports positive-charged particles toward the direction of the magnetic field, while negative-charged particles toward the opposite direction. Similarly for events with more LH particles, i.e., a negative $\mu_5 < 0$, the CME current $\mathbf{J}$ flips its direction, thus transporting positive-/negative-charged particles oppositely to the $\mu_5 > 0$ case.

Such CME-induced charge transport will lead to accumulation of excessive positive-/negative-charges above and below the Reaction Plane, i.e., a charge dipole moment along the out-of-plane (or equivalently the $\mathbf{B}$ field) direction. This charge separation effect, when carried by the bulk collective flow, will lead to a specific pattern of the azimuthal distribution of the finally observed charged hadrons particles in the momentum space, as follows:

$$\frac{dN^{\pm}}{d\phi} \propto 1 + 2a_1^{\pm} \sin(\phi - \Psi_{RP}) + 2v_2 \cos(2\phi - 2\Psi_{RP}) + \dots . \qquad (2.3)$$

Here, Ψ_{RP} represents the azimuthal angle of Reaction Plane, coefficient v_2 is the elliptic flow, while the CME charge separation is represented by the dipole term $a_1^{\pm} \sin(\phi - \Psi_{RP})$ with $a_1^+ = -a_1^-$.

Note however that the direction of the CME current flips with the chirality imbalance (arising from fluctuations), and there are equal probabilities for the event-wise chirality imbalance to be positive or negative. Therefore the event-averaged measurement of $a_1^{\pm}$ is expected to be vanishing, i.e., $\langle a_1^{\pm} \rangle = 0$. Clearly one can only measure the variance of such charged dipole. This can be done by measuring the azimuthal correlations for same-charge and opposite-charge hadron pairs:

$$\gamma_{\alpha\beta} \equiv \langle \cos(\phi_i + \phi_j - 2\Psi_{RP}) \rangle_{\alpha\beta} , \qquad \delta_{\alpha\beta} \equiv \langle \cos(\phi_i - \phi_j) \rangle_{\alpha\beta} , \qquad (2.4)$$

where $\alpha, \beta = +$ or $-$ represents positive-/negative-charged particles. In the absence of *"true"* two-particle correlations, i.e., with the two-particle distribution of the form:

$$f(\phi_i, \phi_j) \equiv f(\phi_i)f(\phi_j) + C(\phi_i, \phi_j) \to f(\phi_i)f(\phi_j), \qquad (2.5)$$

one can derive that

$$\gamma_{\alpha\beta}^{\text{CME}} = -\langle a_{1,\alpha} a_{1,\beta} \rangle, \qquad \delta_{\alpha\beta}^{\text{CME}} = \langle a_{1,\alpha} a_{1,\beta} \rangle. \tag{2.6}$$

It thus seems that by measuring $\gamma_{++,--}$ versus γ_{+-}, one could extract the CME-induced signal. This is however naive, and turns out not working well due to the presence of substantial background correlations in the neglected $C(\phi_i, \phi_j)$ term in the above. Indeed a number of analyses clearly demonstrated that the correlators γ, δ are strongly influenced by non-CME, flow-driven background contributions, such as the local charge conservation [29, 30] and the transverse momentum conservation [31–33]. If one adopts a two-component model analysis as developed in [26, 34], then the correlators could be decomposed into the background contribution F, and the "pure" CME signal $H \equiv \langle a_{1,\alpha} a_{1,\beta} \rangle$. They contribute differently into the γ and δ correlators, and in particular the flow-driven background would contribute to the correlators as $\delta^{\text{bkg}} = F$ and $\gamma^{\text{bkg}} = \kappa v_2 F$. We therefore obtain the following decomposition relations:

$$\gamma = \kappa v_2 F - H, \qquad \delta = F + H. \tag{2.7}$$

Here the factor κ quantifies the amount of v_2-driven background, which is expected to be in the range of 1–1.5 and the AMPT simulation gives the expectation that $\kappa \sim 1.2$ (see, e.g., [35]).

In this dissertation, we will first (see Chap. 4) focus on event-averaged smooth hydro simulations for quantifying the CME-induced signals $H_{SS} \equiv H_{++/--} = (a_1^{ch})^2$ and $H_{OS} \equiv H_{+-} = -(a_1^{ch})^2$, to be compared with the extracted signals by STAR Collaboration. Then in Chap. 5, we will investigate γ and δ correlators with event-by-event AVFD simulations which provides more realistic two-particle correlations by including fluctuations, hadronic cascades as well as various background contributions.

References

1. S.L. Adler, Phys. Rev. **177**, 2426 (1969). https://doi.org/10.1103/PhysRev.177.2426
2. J.S. Bell, R. Jackiw, Nuovo Cim. **A60**, 47 (1969). https://doi.org/10.1007/BF02823296
3. A. Vilenkin, Phys. Rev. **D22**, 3080 (1980). https://doi.org/10.1103/PhysRevD.22.3080
4. D. Kharzeev, Phys. Lett. **B633**, 260 (2006). https://doi.org/10.1016/j.physletb.2005.11.075
5. D. Kharzeev, A. Zhitnitsky, Nucl. Phys. **A797**, 67 (2007). https://doi.org/10.1016/j.nuclphysa. 2007.10.001
6. D.E. Kharzeev, L.D. McLerran, H.J. Warringa, Nucl. Phys. **A803**, 227 (2008). https://doi.org/ 10.1016/j.nuclphysa.2008.02.298
7. K. Fukushima, D.E. Kharzeev, H.J. Warringa, Phys. Rev. **D78**, 074033 (2008). https://doi.org/ 10.1103/PhysRevD.78.074033
8. D.E. Kharzeev, H.U. Yee, Phys. Rev. **D84**, 045025 (2011). https://doi.org/10.1103/PhysRevD. 84.045025

9. T. Schäfer, E.V. Shuryak, Rev. Mod. Phys. **70**, 323 (1998). https://doi.org/10.1103/RevModPhys.70.323

10. D. Diakonov, Nucl. Phys. Proc. Suppl. **195**, 5 (2009). https://doi.org/10.1016/j.nuclphysbps.2009.10.010

11. H.B. Nielsen, M. Ninomiya, Phys. Lett. **130B**, 389 (1983). https://doi.org/10.1016/0370-2693(83)91529-0

12. D.T. Son, B.Z. Spivak, Phys. Rev. **B88**, 104412 (2013). https://doi.org/10.1103/PhysRevB.88.104412

13. Q. Li, D.E. Kharzeev, C. Zhang, Y. Huang, I. Pletikosic, A.V. Fedorov, R.D. Zhong, J.A. Schneeloch, G.D. Gu, T. Valla, Nature Phys. **12**, 550 (2016). https://doi.org/10.1038/nphys3648

14. J. Xiong, S.K. Kushwaha, T. Liang, J.W. Krizan, W. Wang, R.J. Cava, N.P. Ong, in *APS March Meeting 2015 San Antonio, TX, March 2–6, 2015* (2015). http://inspirehep.net/record/1357083/files/arXiv:1503.08179.pdf

15. X. Huang et al., Phys. Rev. **X5**(3), 031023 (2015). https://doi.org/10.1103/PhysRevX.5.031023

16. F. Arnold et al., Nature Commun. **7**, 1615 (2016). https://doi.org/10.1038/ncomms11615

17. B.I. Abelev et al., Phys. Rev. Lett. **103**, 251601 (2009). https://doi.org/10.1103/PhysRevLett.103.251601

18. B.I. Abelev et al., Phys. Rev. **C81**, 054908 (2010). https://doi.org/10.1103/PhysRevC.81.054908

19. L. Adamczyk et al., Phys. Rev. **C88**(6), 064911 (2013). https://doi.org/10.1103/PhysRevC.88.064911

20. L. Adamczyk et al., Phys. Rev. **C89**(4), 044908 (2014). https://doi.org/10.1103/PhysRevC.89.044908

21. L. Adamczyk et al., Phys. Rev. Lett. **113**, 052302 (2014). https://doi.org/10.1103/PhysRevLett.113.052302

22. B. Abelev et al., Phys. Rev. Lett. **110**(1), 012301 (2013). https://doi.org/10.1103/PhysRevLett.110.012301

23. V. Khachatryan et al., Phys. Rev. Lett. **118**(12), 122301 (2017). https://doi.org/10.1103/PhysRevLett.118.122301

24. D.E. Kharzeev, J. Liao, S.A. Voloshin, G. Wang, Prog. Part. Nucl. Phys. **88**, 1 (2016). https://doi.org/10.1016/j.ppnp.2016.01.001

25. J. Liao, Pramana **84**(5), 901 (2015). https://doi.org/10.1007/s12043-015-0984-x

26. A. Bzdak, V. Koch, J. Liao, Lect. Notes Phys. **871**, 503 (2013). https://doi.org/10.1007/978-3-642-37305-3_19

27. X.G. Huang, Rept. Prog. Phys. **79**(7), 076302 (2016). https://doi.org/10.1088/0034-4885/79/7/076302

28. Y. Jiang, S. Shi, Y. Yin, J. Liao, Chin. Phys. **C42**(1), 011001 (2018). https://doi.org/10.1088/1674-1137/42/1/011001

29. S. Schlichting, S. Pratt, Phys. Rev. **C83**, 014913 (2011). https://doi.org/10.1103/PhysRevC.83.014913

30. S. Pratt, S. Schlichting, S. Gavin, Phys. Rev. **C84**, 024909 (2011). https://doi.org/10.1103/PhysRevC.84.024909

31. A. Bzdak, V. Koch, J. Liao, Phys. Rev. **C81**, 031901 (2010). https://doi.org/10.1103/PhysRevC.81.031901

32. A. Bzdak, V. Koch, J. Liao, Phys. Rev. **C83**, 014905 (2011). https://doi.org/10.1103/PhysRevC.83.014905

33. J. Liao, V. Koch, A. Bzdak, Phys. Rev. **C82**, 054902 (2010). https://doi.org/10.1103/PhysRevC.82.054902

34. J. Bloczynski, X.G. Huang, X. Zhang, J. Liao, Nucl. Phys. **A939**, 85 (2015). https://doi.org/10.1016/j.nuclphysa.2015.03.012

35. F. Wen, J. Bryon, L. Wen, G. Wang, Chin. Phys. **C42**(1), 014001 (2018). https://doi.org/10.1088/1674-1137/42/1/014001

Chapter 3
The Anomalous-Viscous Fluid Dynamics Framework

The environment created in a high-energy heavy-ion collision is not a static medium but dominantly an almost-neutral hot fireball undergoing a strong collective expansion. To quantitatively study the signal of Chiral Magnetic Effect, one needs to properly describe the transport of the light fermions (as RH and LH particles) in the hydrodynamic framework and to account for anomaly. Here we adopt a linearization approach to treat the RH and LH fermion currents as perturbations on top of the expanding bulk matter, and describe the evolution by the conservation equations with axial sources:

$$\hat{D}_\mu J_{f,R}^\mu = +\frac{N_c Q_f^2}{4\pi^2} E_\mu B^\mu, \tag{3.1}$$

$$\hat{D}_\mu J_{f,L}^\mu = -\frac{N_c Q_f^2}{4\pi^2} E_\mu B^\mu. \tag{3.2}$$

These equations are solved as a linear perturbation on top of the neutral viscous fluid background, described by data validated boost-invariant VISH2+1 hydrodynamic simulation [1]. The VISH2+1 provides an excellent description of the bulk evolution and the computed collective flow observables agree well with a large body of available data. Adopted from VISH2+1 according to the boost-invariant nature, we employ the coordinates (τ, x, y, η), converting the Minkowski spacetime coordinates $z - t$ into proper time $\tau \equiv \sqrt{t^2 - z^2}$ and rapidity $\eta \equiv \frac{1}{2} \ln \frac{t+z}{t-z}$, with metric convention $g_{\mu\nu} = (1, -1, -1, -\tau^2)$. We denote the projection operator $\Delta^{\mu\nu} = (g^{\mu\nu} - u^\mu u^\nu)$ where u^μ is the fluid velocity field, while $\hat{d} = u^\mu \hat{D}_\mu$ with $\hat{D}_\mu$ as the covariant derivatives. Regardless of absence of gravitational curvature, this coordinate system is subjected to nonzero affine connections $\Gamma_{\mu\nu}^\rho = \frac{1}{2} g^{\rho\sigma} (\partial_\nu g_{\sigma\mu} + \partial_\mu g_{\sigma\nu} - \partial_\sigma g_{\mu\nu})$, and the covariant derivative acting on a Lorentz scaler S, vector V^μ, and tensor $W^{\mu\nu}$ can be expressed as

© Springer Nature Switzerland AG 2019
S. Shi, *Soft and Hard Probes of QCD Topological Structures in Relativistic Heavy-Ion Collisions*, Springer Theses, https://doi.org/10.1007/978-3-030-25482-7_3

$$\hat{D}_\mu S = \partial_\mu S, \quad \hat{D}_\mu V^\nu = \partial_\mu V^\nu + \Gamma^\nu_{\lambda\mu} V^\lambda, \quad \hat{D}_\mu W^{\nu\rho} = \partial_\mu W^{\nu\rho} + \Gamma^\nu_{\lambda\mu} W^{\lambda\rho} + \Gamma^\rho_{\lambda\mu} W^{\nu\lambda}.$$
$$(3.3)$$

It is worth mentioning that when including nonzero vector/axial charge $n = \lambda s$, one could expect that thermal quantities like ϵ, p, s, and T would be modified by $\sim\lambda^2$. The influence on hydro background from the back reaction of nonzero net charge densities is rather small for 200 GeV collisions, and thus the linearized approach here provides a very good approximate description. However, the influence of finite charge becomes crucial for collisions at low beam energy, where net vector/axial charges become substantial. To study anomalous transport in low energy collisions, one should in principle solve the full dynamic equations coupling current transport (Eqs. (3.1) and (3.2)) with evolution of the energy-momentum tensor.

3.1 Viscous Fluid Dynamical Description of Heavy-Ion Collisions

VISH2+1 hydro package [1] is an open source hydrodynamics code package developed by the Ohio State University nuclear theory group. It describes the evolution of the QGP, created by Relativistic Heavy Ion Collisions, taking the assumption that the system to be boost-invariant and charge neutral. Both assumptions are valid for ultra-Relativistic Heavy Ion Collisions (e.g., for top energy collisions at RHIC as well as collisions at the LHC). It adopts the Israel–Steward framework for the second-order viscous hydrodynamic equations, with the fluid energy-momentum tensor given by:

$$T^{\mu\nu} = \varepsilon\, u^\mu u^\nu - (p + \Pi)\, \Delta^{\mu\nu} + \pi^{\mu\nu}. \tag{3.4}$$

VISH2+1 hydro implements the lattice-based equation of state s95p-v0-PCE [2], with the neutrality assumption $n_f \equiv 0$, and solves the energy-momentum conservation equation

$$\hat{D}_\nu T^{\mu\nu} = 0, \tag{3.5}$$

with the relaxation equation of the bulk pressure Π and shear stress tensor $\pi^{\mu\nu}$ toward the corresponding Navier–Stokes form,

$$\Delta^{\mu\alpha}\Delta^{\nu\beta}\hat{d}\pi_{\alpha\beta} = -\frac{1}{\tau_\pi}(\pi^{\mu\nu} - 2\eta\sigma^{\mu\nu}) - \frac{\pi^{\mu\nu}}{2}\frac{\eta T}{\tau_\pi}\hat{D}_\lambda\left(\frac{\tau_\pi}{\eta T}u^\lambda\right) \tag{3.6}$$

$$\hat{d}\Pi = -\frac{1}{\tau_\Pi}(\Pi + \zeta\theta) - \frac{\Pi}{2}\frac{\zeta T}{\tau_\Pi}\hat{D}_\lambda\left(\frac{\tau_\Pi}{\zeta T}u^\lambda\right). \tag{3.7}$$

With appropriate initial conditions, one can then obtain the space-time configuration of the temperature, energy density, pressure, as well as the fluid velocity of the

QGP by solving the above differential equations. At the end of the fluid evolution, the QGP hadronizes at a specific temperature, the freeze-out temperature T_f, and the final hadrons are then locally produced in all fluid cells on the freeze-out hyper-surface according to a local thermal-equilibrium distribution with viscous corrections (see e.g. [1] for details), following the Cooper–Frye freeze-out formula

$$E\frac{dN}{d^3p}(x^\mu, p^\mu) = \frac{g}{(2\pi)^3}\int_{\Sigma_{fo}} p^\mu d^3\sigma_\mu f(x, p). \tag{3.8}$$

It is worth mentioning that during the hadronic stage, the hadron scatterings and resonance decay processes would modify the finally observed hadronic spectra. This has been properly implemented in the VISH2+1 simulations and the results provide good agreement with measurements of identified hadron observables. Finally we emphasize that all these choices are following the standard VISH2+1 setup which has been successfully validated with various experimental data for soft bulk observables.

3.2 Fermion Currents and Anomalous Chiral Transport

Let us focus on the collision with relatively high beam energy, e.g., the top energy collision at RHIC with $\sqrt{s_{NN}} = 200\,\text{GeV}$. The matter produced at such energy has rather small net conserved charges, compared to the bulk energy or entropy, and its evolution is usually well described by viscous hydrodynamics assuming neutrality for all fermion currents. However in order to describe the charge transport in QGP, one needs to include the corresponding fluid dynamical evolution for the fermions (i.e., quarks and antiquarks which carry all the conserved charges). These currents though could be treated in a perturbative way, i.e. by evolving them on top of the neutral bulk fluid background (as specified by the space-time dependent temperature field $T(x^\mu)$ and fluid velocity field $u^\nu(x^\mu)$ from solving Eq. (3.5)) and ignoring their back reaction to the bulk evolution. Furthermore one would need to implement the anomalous transport effect in the fluid dynamics framework [3]. The corresponding fluid dynamical equations for the evolution of both RH and LH fermion currents for each light flavor of quarks, take the following form:

$$\hat{D}_\mu J^\mu_{\chi,f} = \chi\frac{N_c Q^2_f}{4\pi^2}E_\mu B^\mu \tag{3.9}$$

$$J^\mu_{\chi,f} = n_{\chi,f}u^\mu + v^\mu_{\chi,f} + \chi\frac{N_c Q_f}{4\pi^2}\mu_{\chi,f}B^\mu \tag{3.10}$$

$$\Delta^\mu_\nu\hat{d}\left(v^\nu_{\chi,f}\right) = -\frac{1}{\tau_r}\left[\left(v^\mu_{\chi,f}\right) - \left(v^\mu_{\chi,f}\right)_{NS}\right] \tag{3.11}$$

$$\left(v^\mu_{\chi,f}\right)_{NS} = \frac{\sigma}{2}T\Delta^{\mu\nu}\partial_\nu\left(\frac{\mu_{\chi,f}}{T}\right) + \frac{\sigma}{2}Q_f E^\mu \tag{3.12}$$

where $\chi = \pm 1$ labels chirality for RH/LH currents and $f = u, d$ labels different light quark flavors with their respective electric charge Q_f and with color factor $N_c = 3$. The $E^\mu = F^{\mu\nu} u_\nu$ and $B^\mu = \frac{1}{2}\epsilon^{\mu\nu\alpha\beta} u_\nu F_{\alpha\beta}$ represent the external electromagnetic fields in fluid's local rest frame. Furthermore the (small) fermion densities $n_{\chi,f}$ and corresponding chemical potential $\mu_{\chi,f}$ are related by lattice-computed quark number susceptibilities $c_2^f(T)$[4]. It is worth emphasizing that the above framework treats the normal viscous currents $v_{\chi,f}^\mu$ at the second-order of gradient expansion by incorporating relaxation toward Navier–Stokes form (which is the first-order gradient term), thus in consistency with the background bulk flow which is also described by the second-order viscous hydrodynamics as shown in Eqs. (3.5)–(3.7). Two key transport coefficients (characterizing normal viscous effects) are explicitly involved: the normal diffusion coefficient σ and the relaxation time τ_r.

In particular, it should be emphasized that the term $\chi \frac{N_c Q_f}{4\pi^2} \mu_{\chi,f} B^\mu$ in the Eq. (3.10) implements explicitly the CME current. Its sign changes with the chirality χ, which reflects the feature of anomalous transport where the direction of the CME current is opposite for RH and LH particles. Note also that in the above framework, the CME current is treated as an instantaneous term without any second-order thermal relaxation effect, owing to the argument that the CME current is of quantum nature. This is still an open question and a certain fraction of the CME current may still suffer from relaxation effect (see e.g. recent discussions in [5, 6]). In Sect. 4.7 we will introduce the second-order relaxation term for the CME current in a phenomenological way and investigate the potential influence on the CME signal by such relaxation effect.

By solving these equations with proper initial conditions, one can obtain detailed information on the space-time evolution of the fermion currents of any flavor or chirality which accounts for both normal viscous charge transport and the anomalous transport. Then, the relevant chemical potential, for each species of hadrons, can be determined on the hydrodynamic freeze-out hyper-surface and include such nonzero chemical potential in the Cooper–Frye formula (3.8). After including necessary hadron cascade processes, especially the contributions of resonance decay, one can obtain the momentum distribution of the hadrons which would be eventually measured by detectors.

3.3 Comparison of Normal and Anomalous Transport

To illustrate how the charge separation arises from the CME-induced anomalous transport within the AVFD framework, let us visualize how the fermion densities evolve under normal and anomalous transport in Fig. 3.1. When the hydrodynamic evolution starts (at proper time $\tau = \tau_{0,hydro} = 0.6$ fm/c), we initialize the RH/LH u-quark number density as a symmetric one (shown in the left most panel). If there is no external magnetic field applied, i.e., only normal transport, both RH

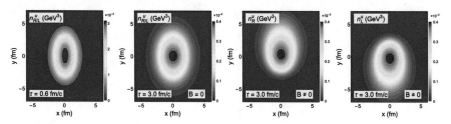

Fig. 3.1 The evolution of u-flavor densities via solving AVFD equations from the same initial charge density distribution (for either RH or LH) at $\tau = 0.60$ fm/c (left most panel) in three cases: (a) (second left panel) for either RH or LH density at $\tau = 3.00$ fm/c with magnetic field $B \to 0$ i.e. no anomalous transport; (b) (second right panel) for RH density and (c) (most right panel) for LH density, both at $\tau = 3.00$ fm/c with nonzero B field along positive y-axis

and LH u-quarks expand with the fluid and also experience viscous transport like diffusion, in a symmetric fashion along x-/y-direction (shown in the second left panel). On the other hand, once an external magnetic field is turned on along the out-of-plane direction $\hat{y}$, the anomalous CME current propagates RH u-quarks toward the direction of B field and LH u-quarks toward the opposite direction, leading to an asymmetric pattern of the charge distribution along the out-of-plane direction (shown in the two right panels).

As a result of the anomalous transport under the presence of chirality imbalance (i.e., either the RH or LH pattern in Fig. 3.1 would dominate), there will be accumulation of opposite charges on the two poles above and below the Reaction Plane. This would therefore lead to a dipole term in the azimuthal distribution of the electric charge chemical potential, $\mu_Q \propto [1 + 2a_1^{ch} \sin(\phi - \Psi_{RP})]$. In Fig. 3.2 we show the dipole coefficient of the electric chemical potential computed on the freeze-out surface at different proper time τ

$$\epsilon_1^{\mu_Q/T}(\tau) \equiv \frac{1}{2\pi T_{\text{dec}}} \int_{T(\tau,\rho,\phi)\equiv T_{\text{dec}}} \mu_Q(\tau, \rho, \phi) \sin(\phi - \Psi_{RP}) \, d\phi. \qquad (3.13)$$

One can see that such dipole coefficient grows as the accumulation of the CME current. At the later stage—after the magnetic field vanishes—the electric dipole eventually gets diluted due to diffusion effect as well as expansion of the bulk background. Upon hadronization via the Cooper–Frye formula (3.8), the dipole term of the chemical potential is converted to the CME-induced charge separation as in Eq. (2.3).

However, the sign of the CME dipole flips event-by-event, according to the sign of chirality imbalance arising from fluctuations in each specific event. Therefore one can only measure this dipole through charge-dependent particle correlations as already discussed before. Nevertheless to quantify the charge separation from CME alone, it suffices to compute the signal with definitive sign of the chirality imbalance, i.e., assuming always more RH particles than LH particles with positive μ_5.

Fig. 3.2 Dipole coefficient of electric chemical potential at freeze-out hyper-surface. Values of magnetic field and chirality imbalance at different centrality range can be found in Sect. 4.5

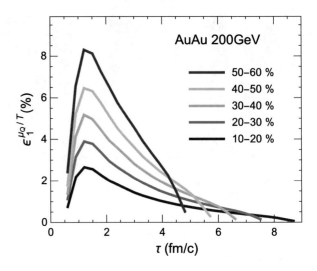

For events with negative μ_5, the charge separation a_1 changes its sign accordingly but the contribution to the two-particle correlations would be the same as the events with positive μ_5.

In passing, it is worth noting that there have been a number of early attempts in applying the anomalous hydrodynamic framework toward describing charge transport in heavy-ion collisions [7–10]. The AVFD framework developed in [11, 12] is by far the most matured approach for state-of-the-art simulations of anomalous transport in heavy-ion collisions based on realistic bulk evolution and incorporating normal viscous transport effects simultaneously. Such a framework allows a quantitative understanding of the generation and evolution of CME-induced charge separation signal in the hydro evolution stage, together with its dependence on various ingredients such as the initial conditions, the magnetic fields, as well as the viscous transport coefficients, as will be reported with great detail in Chap. 4, on top of smooth bulk backgrounds. While in Chap. 5, we will further discuss how event-by-event fluctuating bulk background would influence CME observables.

References

1. C. Shen, Z. Qiu, H. Song, J. Bernhard, S. Bass, U. Heinz, Comput. Phys. Commun. **199**, 61 (2016). https://doi.org/10.1016/j.cpc.2015.08.039
2. P. Huovinen, P. Petreczky, Nucl. Phys. **A837**, 26 (2010). https://doi.org/10.1016/j.nuclphysa. 2010.02.015
3. D.T. Son, P. Surowka, Phys. Rev. Lett. **103**, 191601 (2009). https://doi.org/10.1103/ PhysRevLett.103.191601
4. S. Borsanyi, Z. Fodor, S.D. Katz, S. Krieg, C. Ratti, K. Szabo, J. High. Energy Phys. **01**, 138 (2012). https://doi.org/10.1007/JHEP01(2012)138

5. D.E. Kharzeev, M.A. Stephanov, H.U. Yee, Phys. Rev. **D95**(5), 051901 (2017). https://doi. org/10.1103/PhysRevD.95.051901
6. A. Huang, Y. Jiang, S. Shi, J. Liao, P. Zhuang, Phys. Lett. **B777**, 177 (2018). https://doi.org/ 10.1016/j.physletb.2017.12.025
7. Y. Yin, J. Liao, Phys. Lett. **B756**, 42 (2016). https://doi.org/10.1016/j.physletb.2016.02.065
8. Y. Hirono, T. Hirano, D.E. Kharzeev (2014). arXiv:1412.0311
9. H.U. Yee, Y. Yin, Phys. Rev. **C89**(4), 044909 (2014). https://doi.org/10.1103/PhysRevC.89. 044909
10. M. Hongo, Y. Hirono, T. Hirano, Phys. Lett. **B775**, 266 (2017). https://doi.org/10.1016/j. physletb.2017.10.028
11. Y. Jiang, S. Shi, Y. Yin, J. Liao, Chin. Phys. **C42**(1), 011001 (2018). https://doi.org/10.1088/ 1674-1137/42/1/011001
12. S. Shi, Y. Jiang, E. Lilleskov, J. Liao, Annals Phys. **394**, 50 (2018). https://doi.org/10.1016/j. aop.2018.04.026

Chapter 4
Quantitative Study of the CME Signal

With the AVFD framework, we are now ready to explore the quantitative aspects of the CME-induced charge separation effect in heavy-ion collisions. There are a number of important model inputs: first, the magnetic field strength as well as its time dependence; second, the initial conditions for the fermion charge densities; lastly, the two viscous transport parameters namely the diffusion coefficient as well as the relaxation time. In addition there is a hadronic re-scattering stage after the hydrodynamic freeze-out, the influence of which needs to be understood. In this section, we will systematically investigate the influences of all these factors and discuss the corresponding theoretical uncertainties. Finally, based on our best choices for such model input, we will quantitatively compute the CME-induced H-correlator and compare the results with available experimental data.

4.1 Influence of the Magnetic Field

As the "driving force" of the CME current, the strength and space-time dependence of the magnetic field are among the most crucial factors in quantifying the CME signal. In a heavy-ion collision the two colliding nuclei are positively charged and move at nearly the speed of light, thus producing extremely large magnetic fields in the collision zone. For example, the peak value of such $\mathbf{B}$ field reaches as high as $eB \sim 5m_\pi^2$ (or $B \sim 10^{15}$ T) at $\sqrt{s_{NN}} = 200$ GeV collisions at RHIC, and $\sim 70m_\pi^2$ at $\sqrt{s_{NN}} = 2.76$ TeV LHC collisions.

Many calculations have been done to quantify the magnetic field. For example, by using event-by-event simulations with Mont-Carlo Glauber model, the peak value of the magnetic field is well determined, and its azimuthal orientation with respect to event-wise bulk geometry has been quantified to be roughly along out-of-plane direction with a de-correlation factor [1]. Both its magnitude and its

© Springer Nature Switzerland AG 2019
S. Shi, *Soft and Hard Probes of QCD Topological Structures
in Relativistic Heavy-Ion Collisions*, Springer Theses,
https://doi.org/10.1007/978-3-030-25482-7_4

direction are also found to vary only very mildly in the collision overlapping zone except near the fireball edge. Nevertheless, the **B** time evolution remains an open question. The main source of the magnetic field, namely the spectator nucleons, pass through each other and fly away quickly from the collision zone at mid rapidity. As a result, the field strength from such external sources decays rapidly with a behavior roughly following the formula $B(\tau) = B(0)/(1 + \tau/\tau_B)^{3/2}$, with $\tau_B \sim R_{\text{nuclei}}/\gamma$. On the other hand, the hot medium created in the collision is a conducting plasma and therefore could in principle delay the decrease of the magnetic field through the generation of an induction current in response to the changing magnetic field. To fully address this issue, one needs to treat both the medium and the magnetic field as dynamically evolving together. While many efforts have been made to compute the time dependence of the magnetic field [2–5], the answers from different studies vary considerably. To get an idea of the current status, we show in Fig. 4.1 (left panel) a comparison of various results for the time dependence of the magnetic field: the study by McLerran-Skokov [2] with conductivity $\sigma = \sigma_{\text{LQCD}}$, $10^2\sigma_{\text{LQCD}}$, and $10^3\sigma_{\text{LQCD}}$, and ECHO-QGP simulation [5], as well as three types of parameterizations $B \propto (1+\tau^2/\tau_B^2)^{-1}$, $(1+\tau^2/\tau_B^2)^{-3/2}$, and $\exp(-\tau/\tau_B)$. Clearly, the stronger the medium feedback is, the longer the magnetic field lasts.

The lifetime of the magnetic field strength has a direct and significant impact on the anomalous transport and thus the CME-induced charge separation signal in the end. The AVFD tool allows a quantitative calibration on influence of the uncertainty in magnetic field lifetime on the predicted CME signal. In Fig. 4.1 (middle panel), we compare the results for charge separation a_1^{ch} computed with the various different choices of the magnetic field time dependence. Note all these calculations are done with the same initial axial charge condition $n_5/s = 0.1$ and with the same peak value of **B** field at time $\tau = 0$. As a consequence of the huge difference in the **B** field "surviving" time, the obtained charge separation signal varies substantially across various time dependence schemes. For the three different **B** parameterizations we further compare them, in Fig. 4.1 (right panel), by showing the a_1^{ch} versus the magnetic field lifetime parameter τ_B. Clearly the CME signal grows rapidly with τ_B in all three cases while for the same lifetime τ_B these three parameterizations still show visible difference.

At present there is no convincing conclusion yet regarding the exact time dependence (perhaps except those over-optimistic scenarios), but the comparative study provides a good idea of the associated uncertainties. In the rest part of this chapter, we will use a relatively intermediate case of the following parameterization:

$$\mathbf{B} = \frac{B_0}{1 + \tau^2/\tau_B^2}\hat{y} \tag{4.1}$$

and we adopt a rather "conservative" choice that the lifetime of the magnetic field is comparable to the starting time of the hydrodynamic evolution $\tau_{0,\text{hydro}}$, i.e., with $\tau_B = \tau_{0,\text{hydro}} = 0.6$ fm/c (as represented by the red dots in Fig. 4.1 (middle and

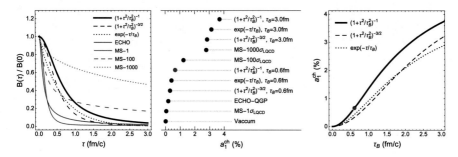

Fig. 4.1 (left) The time dependence of the magnetic field from different study: ECHO-QGP [5] (blue curve), McLerran-Skokov [2] with different electric conductivity (solid red curve for $\sigma = \sigma_{LQCD}$, dashed red line for $\sigma = 100\sigma_{LQCD}$, and dotted red line for $\sigma = 1000\sigma_{LQCD}$). Also, the thick, dashed, and dotted black curves represent the formulations $B \propto (1 + \tau/\tau_B)^{-1}$, $(1 + \tau^2/\tau_B^2)^{-3/2}$, and $\exp(-\tau^2/\tau_B^2)$ respectively, with $\tau_B = 0.6$ fm/c. (middle) Comparison of charge separation predicted by different time dependences of magnetic field. (right) Charge separation a_1 predicted by different magnetic field lifetime τ_B. The thick, dashed, and dotted black curves correspond to the formulations $B \propto (1 + \tau^2/\tau_B^2)^{-1}$, $(1 + \tau^2/\tau_B^2)^{-3/2}$, and $\exp(-\tau/\tau_B)$. In both middle and right panels, the red dots correspond to the magnetic field time dependence to be used for the rest of this study

right panels). The peak value of the magnetic field B_0 for each centrality is taken from the event-by-event Monte-Carlo simulations where the field value is already projected along the elliptic Event Plane to account for the angular de-correlation between field direction and bulk geometry due to fluctuations [1].

4.2 Influence of the Initial Conditions

Another key ingredient of the CME-induced transport is the chirality imbalance, or more generally speaking, the initial conditions of RH/LH charge densities (or equivalently the initial conditions for the vector and axial charges of each flavor of fermions). In this section, we study the influence of initial conditions of charge distribution on the final hadron charge separation with the AVFD tool. For this framework, one needs to provide the following initial conditions, namely the initial four-current $J^\mu_{\chi,f}(\tau = \tau_0)$ for each flavor f as well as chirality χ. For most part of this study, we use null initial condition for the spatial three-current components, i.e., setting all $\mathbf{J}_{\chi,f}(\tau_0) \to 0$ (except in Sect. 4.9) while only consider the zeroth component, i.e., the number densities $J^0_{\chi,f}(\tau = \tau_0)$. Equivalently one can set for each flavor the initial vector and axial charge densities $n = J^0_R + J^0_L$ and $n_5 = J^0_R - J^0_L$. In the following we discuss the influence of the vector and axial charge initial conditions, respectively.

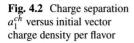

Fig. 4.2 Charge separation a_1^{ch} versus initial vector charge density per flavor

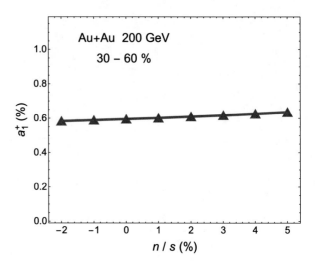

4.2.1   Vector Charge Initial Conditions

As demonstrated earlier, it is the (vector) chemical potential at the freeze-out hypersurface that directly affects the observed particle yields as well as the CME-induced charge separation a_1. Such chemical potential is determined from the corresponding vector number density, the number density of all species of quarks including both RH and LH sectors. One would be interested in how the initial conditions of vector charge affects the CME signal. As a test, we take the initial condition for the vector charge densities to be proportional to the entropy density at the hydro initial time, and then vary this proportionality coefficient to examine its effect on the final charge separation a_1^{ch}. In Fig. 4.2 we show the a_1^{ch} computed from AVFD with a wide range of different initial vector charge densities (but with the same fixed initial axial charge density $n_5/s = 0.1$). The resulting signal stays roughly constant despite even significant changes of the vector densities. Clearly this suggests an insensitivity of the CME signal to the initial vector charge densities, which would therefore imply a negligible potential influence from the uncertainty in constraining the initial conditions for vector charge densities. For most AVFD simulations performed in this study, we set a small but nonzero initial vector charge density as $n_{u,d}/s = 1\%$ (due to stopping) which is a very reasonable estimate for top energy collisions at RHIC as also indicated by e.g. AMPT simulations.

4.2.2   Axial Charge Initial Conditions

The axial charge quantifies the number difference between RH and LH fermions. As discussed before, it plays a key role in the generation of the CME current and hence

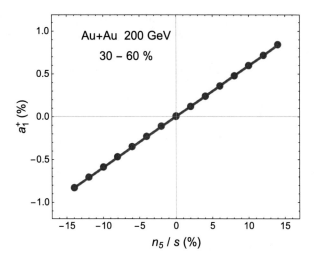

Fig. 4.3 Charge separation a_1 versus initial axial charge density per flavor

the final charge separation signal which is expected to sensitively depend upon the initial condition for the axial charge.

For what we consider, the axial charge density is small compared with entropy density (or equivalently the corresponding chemical potential being small compared with temperature), and therefore the axial charge density and axial chemical potential are linearly proportional to each other. Thus one expects $J^{\mu}_{CME} = C_A \mu_5 B^{\mu} \propto n_5$, i.e., the final charge separation signal should be roughly linearly dependent on the amount of initial axial charge. This linear dependence is indeed verified to be true, as shown in Fig. 4.3. When $n_5 = 0$, there is no chiral imbalance, and no CME-induced charge separation as expected.

As demonstrated above, the CME charge separation is mainly controlled by chirality imbalance while insensitive to the vector charge initial condition, so let us focus on how to properly estimate the axial charge initial condition. Following the scenario based on chirality imbalance arising from gluonic topological charge fluctuations in the early-stage glasma [6–8], the starting point is the following estimation of the axial charge density fluctuations:

$$\sqrt{n_5^2} = \left[\tau_0 \frac{(gE^c)(gB^c)}{16\pi^2} \right] \times \sqrt{N_{tube}} \times \frac{\pi \, \rho_{tube}^2}{A_{overlap}}. \qquad (4.2)$$

The part inside $[\dots]$ counts the density of axial charge n_5 in a *single* glasma flux tube in which the parallel/anti-parallel chromo-electric and chromo-magnetic fields E^c, B^c provide nonzero topological charge density thus nonzero induce corresponding axial charge density via the standard anomaly formula. The total number of glasma flux tubes N_{tube} in a given event can be estimated from the binary collision number, $N_{tube} \simeq N_{coll}$. Note that the color fields inside glasma flux tube are very strong, i.e., $E^c, B^c \sim Q_s^2/g$. However inside each flux tube the E^c, B^c fields randomly take parallel or anti-parallel configurations and thus gives randomly positive or negative contributions to axial charge: this reduces the net axial charge fluctuation one could get in each event, and such reduction effect is taken in account by the $\sqrt{N_{\text{tube}}}$. Finally, to get an averaged/smeared-out axial charge density over the whole fireball transverse area by the contribution of individual flux tube, we include a "dilution" factor $\frac{\pi \rho_{\text{tube}}^2}{A_{\text{overlap}}}$. Here $\pi \rho_{\text{tube}}^2$ is for the flux tube transverse area (with $\rho_{\text{tube}} \simeq 1\text{fm}$ the transverse extension of a glasma flux tube) while A_{overlap} is the transverse geometric overlapping area for the collision zone of the two colliding nuclei. Putting these all together, one then obtains the following estimate:

$$\sqrt{\langle n_5^2 \rangle} \simeq \frac{Q_s^4 \, (\pi \rho_{\text{tube}}^2 \tau_0) \, \sqrt{N_{\text{coll.}}}}{16\pi^2 \, A_{\text{overlap}}}. \qquad (4.3)$$

The above estimate is then used to determine a ratio λ_5 of total average axial charge over the total entropy in the fireball at initial time τ_0, $\lambda_5 \equiv \frac{\int_V \sqrt{\langle n_5^2 \rangle}}{\int_V s}$ where the integration is over fireball spatial volume and the s is the entropy density from bulk hydro initial condition. This ratio is then used in the AVFD simulations to set an initial axial charge distribution locally proportional to entropy density via $n_5^{\text{initial}} = \lambda_5 \, s$. This properly reflects the fact that axial charge arises from local domains with gluon topological fluctuations and that there are more such domains where the matter is denser. Such axial charge density estimate depends most sensitively upon the saturation scale Q_s, in the reasonable range of $Q_s^2 \simeq 1 \sim 1.5 \, \text{GeV}^2$ for RHIC 200 GeV collisions [9, 10].

The axial charge estimated from Eq. (4.8), while not large, is not very small. This is likely due to the out-of-equilibrium nature of the glasma with very strong color fields and large topological fluctuations as compared with usual expectation from perturbative thermal plasma case. The estimate of initial axial charge used in Ref. [8] and here would correspond to a relatively large Chern–Simons diffusion rate $\Gamma_{\text{CS}} \sim Q_s^4$ which is much larger than the usual thermal values. Interestingly this rapid rate has been confirmed recently in Ref. [11] which extracts the rate by using classical-statistical real time lattice simulation in non-equilibrium glasma of weakly coupled but highly occupied gauge fields.

4.3   Dependence on the Viscous Transport Parameters

While the main new interest and recent developments focus on the anomalous transport and its consequence on the charge distribution in QGP, it is quite obvious that more conventional viscous transport like charge diffusion would certainly also affect the charge distribution. However such contributions are largely ignored in past studies and it was unclear to which extent the CME signal would depend on normal viscous transport. The AVFD framework for the first time allows quantitative study of this problem and helps constraining the important theoretical uncertainty due to the viscous transport.

As shown in Eqs. (3.5)–(3.12), the AVFD framework is based on second-order Israel–Stewart and Navier–Stokes equations. Hence it includes the diffusion and conduction effects in the first order as well as the relaxation effect in the second order. Such viscous transport is controlled by two key parameters: the diffusion coefficient and the relaxation time. In this subsection, we quantify how CME signals are influenced by these parameters.

4.3.1 Dependence on the Diffusion Coefficient

Diffusion effect is the macroscopic manifestation of the Brownian motion of the particles. It causes the conserved charge density to spreads out under the presence of density gradient, leading eventually to homogeneous distribution in thermal equilibrium. The diffusion coefficient σ controls how fast the diffusion process transports charges around.

The dependence of the charge separation a_1 on the diffusion coefficient σ, is shown in Fig. 4.4. A large value of σ would imply strong and fast diffusion for any nonzero charge density and therefore would suppress the charge separation induced by CME by transporting net charges across the Reaction Plane. Indeed we see that the signal a_1 decreases with increasing σ when the diffusion effect is strong. On the other hand, when σ is small, the diffusion is not strong enough to bring net charges from one side of the Reaction Plane to the other side. Instead it can help the CME-induced charge dipole spread out a little more over the freeze-out hyper-surface and slightly enhance the a_1 of the final hadrons. One can see that such small diffusion effect indeed slightly increases the charge separation signal.

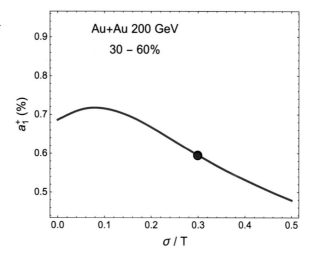

Fig. 4.4 The charge separation a_1 as a function of the diffusion coefficient σ (scaled by the temperature T). The red dot indicates the commonly chosen value, to be used later

Fig. 4.5 The charge
separation a_1 as a function of
the relaxation time τ_r (scaled
by T^{-1}). The red dot
indicates the commonly
chosen value, to be used later

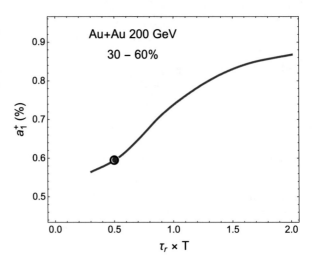

4.3.2 Dependence on the Relaxation Time

Relaxation time parameter τ_r controls the time scale that is needed to build up
the "diffusion current" in response to the density gradient. Intuitively speaking, a
small (large) τ_r implies rapid (slow) building up of the diffusion current and causes
stronger (weaker) diffusion effect, thus suppressing more (less) the CME-induced
charge separation. Indeed, as shown in Fig. 4.5, the signal a_1 increases steadily with
increasing relaxation time.

In summary, one can see that both the diffusion and relaxation effects have
considerable influence on the charge separation signal. A commonly adopted choice
of $\sigma = 0.3T$ and $\tau_r = 0.5/T$ (see e.g. [12]), as indicated by the red blob in
Figs. 4.4 and 4.5, will be used in our later computations. It is worth emphasizing
that the choice of diffusion parameter here corresponds to the electric conductivity
$\sigma_{ele} = e^2\sigma \sim 5\,\text{MeV}$, which is consistent with the conductivity given by lattice
simulations (see e.g. [13]). The curves in Figs. 4.4 and 4.5 shall give a quite clear
idea of the uncertainty in the signal due to the uncertainty associated with the input
values of σ and τ_r.

4.4 Contribution from Resonance Decay

In addition to the viscous transport, another "trivial"/conventional effect which was
often not included properly in previous simulations but which bears quantitative
consequences, is the contribution from resonance decay in the hadronic cascade
stage. As the lightest and most abundant particle, the finally observed pions receive a

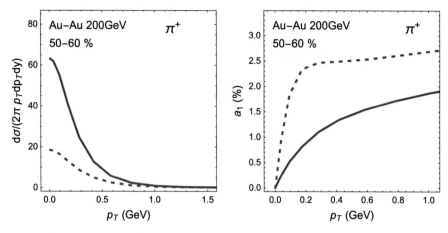

Fig. 4.6 Comparison of differential cross-section (left) and charge separation signal a_1 (right) for π^+ directly produced at the freeze-out surface (dashed curves), versus the final observables including resonance decay contributions (solid curves)

substantial contribution from the feed-down of resonance decays (see Fig. 4.6 (left)), which clearly affect the various bulk observables (usually dominated by pions) such as the harmonic flow coefficients as well as the charge distributions. With the resonance decays already implemented in the VISH2+1 code, the AVFD simulation takes such effect into account and thus allows investigation of their impact on the desired charge separation signal.

In Fig. 4.6 (right) we show the comparison of charge separation a_1 computed from π^+ directly produced at the freeze-out surface, versus that computed from the final observed particles including resonance feed-down. One can see that after taking such effect into account, a_1 is suppressed by $\sim 30\%$. Noting that the main source of decay-produced $\pi^\pm$'s are from the processes like $\rho \to \pi\pi$, $\eta \to \pi\pi\pi$, etc., we take the $\rho^{0,\pm}$ mesons as an example to explain such suppression effect. First of all, the charged $\rho^\pm$ mesons are affected by CME and carry nonzero charge separation a_1 in their distributions. However after their decay into pions by $\rho^\pm \to \pi^0\pi^\pm$, the momentum direction of the parent ρ is not perfectly preserved by the daughter pion, thus smearing out the charge separation initially carried by rho mesons. Secondly, the uncharged ρ^0 mesons do not carry any CME charge separation and neither do the daughter particles from $\rho^0 \to \pi^+\pi^-$, but these decay pions still contribute to the total number of charged pions thus diluting out the observed charge separation signal. In both cases, the charge separation a_1 will be suppressed due to the smearing and dilution, resulting in a significantly reduced magnitude of the signal. This effect must be quantitatively taken into account for meaningful predictions and comparison with experimental data.

4.5   Quantifying the CME Signal

Given the above detailed investigations on the various theoretical inputs and how they influence the charge separation at quantitative level, we now proceed to quantify the CME signal with our best constrained parameter choices. For the magnetic field, we assume them to be homogeneous in space while to evolve in time according to Eq. (4.1), with the lifetime parameter $\tau_B = \tau_{0,\text{hydro}} = 0.6\,\text{fm/c}$. The initial condition for the axial charge is given by Eq. (4.3), with the saturation scale Q_s^2 in a reasonable range of 1–1.5 GeV^2 for the top RHIC energy collisions. The diffusion coefficient and relaxation time are chosen with the commonly used values as $\sigma = 0.3T$, and $\tau_r = 0.5/T$. The resonance decay contributions are properly taken into account. Finally, all the p_T-integrated results in this chapter (for observables like the charge separation a_1 and two-particle correlation H) are computed from hadrons in the range of $0.15 < p_T < 2\,\text{GeV}$, which is exactly the same as the experimental kinematic cuts adopted in the relevant STAR measurements.

The AVFD results for the charge dependent H-correlations, obtained with the aforementioned parameters, for various centrality bins are presented in Fig. 4.7, with the green band spanning the range of key parameter Q_s^2 in the 1–1.5 GeV^2 range to reflect the uncertainty in estimating initial axial charge (see Eq. (4.3)). Clearly the CME-induced correlation is very sensitive to the amount of initial axial charge density as controlled by Q_s^2, especially in the peripheral collisions. The comparison with STAR data [14] shows very good agreement for the magnitude and centrality trend, provided that the Q_s value is at the relatively larger end of the quoted band. In Table 4.1 we show the values of the magnetic field peak strength and the initial axial charge density (normalized by entropy density) that are used in computing the Fig. 4.7.

Fig. 4.7 Quantitative predictions from Anomalous-Viscous Fluid Dynamics simulations for the CME-induced H-correlations, in comparison with STAR measurements [14]. The uncertainty of experimental data comes from the uncertainty of κ. Central values correspond to $\kappa = 1.2$, while upper and lower bounds are for $\kappa = 1$ and $\kappa = 1.5$, respectively. The green bands reflect current theoretical uncertainty in the initial axial charge generated by gluonic field fluctuations

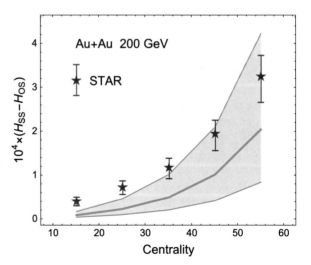

Table 4.1 Centrality dependence of magnetic field peak strength and the initial chirality imbalance

Centrality bin	10–20%	20–30%	30–40%	40–50%	50–60%
$eB_0(m_\pi^2)$	2.34	3.10	3.62	4.01	4.19
n_5/s	0.065	0.078	0.095	0.119	0.155

The n_5/s shown here is obtained with a saturation scale $Q_s^2 = 1.25\,\text{GeV}^2$

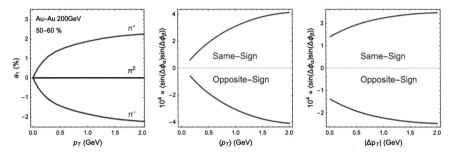

Fig. 4.8 (left) Transverse momentum p_T dependence of charge separation a_1; (middle) Out-of-plane two-particle correlations versus $\langle p_T \rangle \equiv \frac{1}{2}(|p_{T,\alpha}| + |p_{T,\beta}|)$; (right) Out-of-plane two-particle correlations versus $|\Delta p_T| \equiv \left||p_{T,\alpha}| - |p_{T,\beta}|\right|$

The AVFD simulations can further provide p_T-differential information of the CME signal. As shown in Fig. 4.8 left panel (for 50–60% most-central 200 GeV Au–Au collisions), the charge separation a_1 increases with higher and higher transverse momentum, basically following a hydrodynamic pattern via Cooper-Frye thermal production. One can also examine the associated out-of-plane two-particle correlations

$$\langle \sin(\Delta\phi_\alpha) \sin(\Delta\phi_\beta) \rangle \equiv \langle \sin(\phi_\alpha - \psi_{EP}) \sin(\phi_\beta - \psi_{EP}) \rangle_{\alpha,\beta}, \tag{4.4}$$

versus the pair-"averaged" transverse momentum $\langle p_T \rangle$ and the pair-"relative" transverse momentum $|\Delta p_T|$ defined as

$$\langle p_T \rangle \equiv \frac{1}{2}(|p_{T,\alpha}| + |p_{T,\beta}|), \tag{4.5}$$

$$|\Delta p_T| \equiv \left||p_{T,\alpha}| - |p_{T,\beta}|\right|. \tag{4.6}$$

The results for such dependence are shown in Fig. 4.8 middle and right panels (for 50–60% most-central 200 GeV Au–Au collisions). While the $\langle p_T \rangle$-dependence shows a similar trend as the individual p_T, the $|\Delta p_T|$-dependence shows a somewhat flatter trend due to the fact that contributions to fixed $|\Delta p_T|$ come from the whole momentum regime. Both results show qualitative agreement with the experimental measurements [15–17].

4.6 Predictions for Charge Separation in Isobaric Collisions

As previously emphasized, the main challenge for the search of CME in heavy-ion collisions is to separate CME-induced signal from background correlations. The difficulty lies in that many sources contribute as backgrounds and currently these contributions are poorly constrained theoretically. Therefore, it is highly desirable to develop more experimentally oriented approach such as new analysis methods or new observables. Besides the two-component decomposition into H-correlation as discussed above, a number of different proposals were also put forward, see e.g. [18–25], each with certain advantages. A most promising approach, is to conduct a dedicated isobaric collision experiment, which has now been planned for the 2018 run at RHIC [19]. In such "contrast" colliding systems (specifically for $^{96}_{40}\text{Zr}-^{96}_{40}\text{Zr}$ versus $^{96}_{44}\text{Ru}-^{96}_{44}\text{Ru}$), they have the same baryonic number A but different electric charge Z. The expectation is that their bulk evolutions (and thus background correlations) would be basically identical while their magnetic field strength would be different which in turn implies a corresponding difference in the CME signal. For Ru and Zr isobars, there is a $\sim 10\%$ difference in the total charge as well as magnetic field strength; therefore, a shift of $\sim 20\%$ should be expected for CME-driven correlations on top of identical backgrounds between the two. This will be a crucial test for the search of CME, and quantitative predictions are important.

With the AVFD framework developed here for RHIC collisions at $200\,\text{GeV}$, we now quantitatively compute the expected CME signal for the isobar colliding systems. Note that the various inputs for AVFD have been fixed via Au–Au collisions and there is no further tuning of parameters, i.e., we take the initial magnetic field as that projected with respect to the Participant Plane, $B(\tau = 0) \equiv \langle B^2 \cos(2\psi_B - 2\psi_2 - \pi)\rangle^{1/2}$, with lifetime $\tau_B = 0.6\,\text{fm/c}$, and $Q_s^2 = 1.25\,\text{GeV}^2$ as the "central curve." In Fig. 4.9 (upper right panel), we show the AVFD predictions for pure CME signal $H_{\text{SS}} - H_{\text{OS}} = 2(a_1^{ch})^2$ in these two systems. As expected, one can clearly see the $\sim 20\%$ difference between them. It is worth emphasizing that the absolute value of H-correlator is sensitive to the chirality imbalance n_5/s, determined by saturation scale Q_s, the relative difference of H-correlator between Ru–Ru and Zr–Zr systems is independent.

On the other hand, as such "pure" signal H is not measured directly in experiments, one would be interested in also computing the directly measured correlation $\gamma_{\text{OS}} - \gamma_{\text{SS}}$, and examining whether there would be sizable difference in $\gamma_{\text{OS}} - \gamma_{\text{SS}}$ between the isobaric systems. To do so, however, requires knowledge about the non-CME background contributions. Our strategy here is to extrapolate the likely background level to the isobar collisions, based on the $F + H$ two-component decomposition and the assumption that F is mainly a function of multiplicity only [26]. From the results of RHIC $\sqrt{s_{NN}} = 200\,\text{GeV}$ Au–Au collisions [14], one obtains the background $F_{\text{OS}} - F_{\text{SS}}$ versus the measured charged particle multiplicities, shown as blue circles in Fig. 4.9 (lower left panel). We have also performed a fitting of the $F_{\text{OS}} - F_{\text{SS}}$ versus multiplicity (for Au–Au points) with an algebraic-fractional-formula $F(x) = \dfrac{1 + b_1 x + b_2 x^2}{c_1 x + c_2 x^2}$, shown as the blue curve.

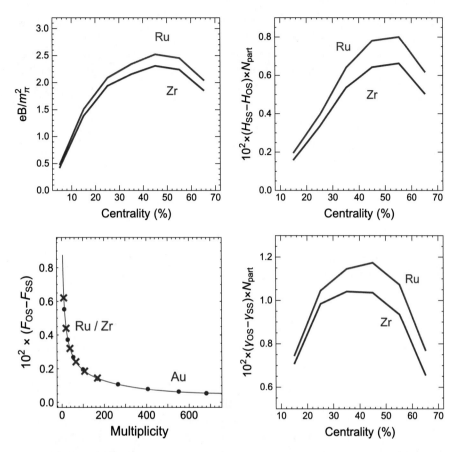

Fig. 4.9 (upper left) Projected initial magnetic field with respect to Participant Plane given by Monte-Carlo Glauber simulation. (upper right) AVFD predictions for CME-induced H-correlations in isobar collisions; (lower left) Fitting of background-induced F-correlations from Au–Au measurement (blue circle), and the corresponding expectations for Zr–Zr and Ru–Ru systems (red cross); (lower right) Predicted γ-correlations in Zr–Zr and Ru–Ru collisions by folding together F- and H-correlations

(The best fit is given by $b_1 = 0.119$, $b_2 = 5.81 \times 10^{-5}$, $c_1 = 33.7$, and $c_2 = 0.395$.) Given the fitting curve, one could then read a plausible estimate of the expected F-correlations in correspondence to the multiplicity in Ru–Ru and Zr–Zr systems, shown as the red crosses in the middle panel of Fig. 4.9 for centrality bins from 10–20% (with higher multiplicity) to 60–70% (with lower multiplicity).

With both the AVFD-predicted CME signal H-correlations and the extrapolated bulk background F-correlations (see Table 4.2), one can then make a prediction for the γ-correlations via Eq. (2.7). The results are shown in Fig. 4.9 (lower right panel), there is a visible ~15% relative difference between the two systems for the γ-correlations in the relatively peripheral collisions. Provided the present

Table 4.2 Centrality dependence of participant number, multiplicity, B field, axial charge, computed CME signal H, and extrapolated non-CME background F in isobar system

Centrality bin	10–20%	20–30%	30–40%	40–50%	50–60%	60–70%
N_{part}	100.46	67.17	43.04	25.79	14.06	6.81
Multiplicity	167.0	106.9	65.43	37.43	19.52	9.096
$eB_0[Ru](m_\pi^2)$	1.507	2.086	2.340	2.520	2.450	2.044
$eB_0[Zr](m_\pi^2)$	1.381	1.937	2.155	2.307	2.239	1.858
n_5/s	0.097	0.114	0.144	0.188	0.266	0.394
$10^4 \times (H_{OS} - H_{SS})[Ru]$	0.196	0.587	1.487	3.023	5.682	9.091
$10^4 \times (H_{OS} - H_{SS})[Zr]$	0.160	0.497	1.245	2.488	4.705	7.425
$10^3 \times (F_{OS} - F_{SS})$	1.42	1.83	2.38	3.18	4.37	6.19

The n_5/s and corresponding H_{OS-SS} correlators shown here are obtained with a saturation scale $Q_s^2 = 1.25\,\mathrm{GeV}^2$

projections for uncertainty and limitations in measurements [19, 20], a 15%-level difference should be readily detectable at the scheduled isobaric collision experiment.

4.7   Possible Relaxation Effect for the Anomalous Transport Currents

As briefly mentioned in Sect. 3.2, the CME current is treated without any relaxation effect in the "standard" AVFD framework, which assumes that the CME current establishes instantaneously in response to the magnetic field and chirality imbalance. In other words, the time needed for the hot medium to respond to the changing external field or axial charge, is assumed to be negligible as compared with relevant QGP evolution time scales. This is plausible as the CME current is a non-dissipative quantum transport current that should occur on microscopic quantum evolution time scale which usually is negligibly short compared with any macroscopic scale. Theoretically this is a subtle issue that has not been fully settled. A recent discussion in [27] suggests that a fraction of the total CME current may be subject to thermal relaxation effect. In such situation, it would be crucial to calibrate the influence of such theoretical uncertainty.

In order to do this, the AVFD framework needs to be slightly adapted to treat a proper fraction of the CME current in a similar fashion as the Navier–Stokes current, i.e., the relaxation effect should be included via the second order viscous terms. In this section, we investigate this uncertainty by comparing two extreme scenarios: the "instant-CME" as in Eqs. (3.9)–(3.12), versus the 100%-relaxation scenario below:

$$\hat{D}_\mu J^\mu_{\chi,f} = \chi \frac{N_c Q_f^2}{4\pi^2} E_\mu B^\mu \tag{4.7}$$

Fig. 4.10 Comparison of two AVFD calculations with different scenarios for the relaxation effect on the CME current: the "instant-CME" scenario (solid green line) without relaxation versus the 100%-relaxation scenario (dashed red line)

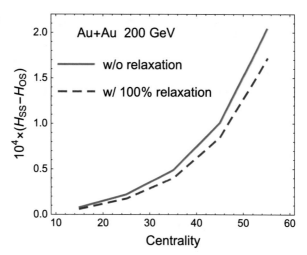

$$J^{\mu}_{\chi,f} = n_{\chi,f}\, u^{\mu} + v^{\mu}_{\chi,f} \tag{4.8}$$

$$\Delta^{\mu}_{\nu}\hat{d}\left(v^{\nu}_{\chi,f}\right) = -\frac{1}{\tau_r}\left[\left(v^{\mu}_{\chi,f}\right) - \left(v^{\mu}_{\chi,f}\right)_{NS}\right] \tag{4.9}$$

$$\left(v^{\mu}_{\chi,f}\right)_{NS} = \frac{\sigma}{2}T\Delta^{\mu\nu}\partial_{\nu}\left(\frac{\mu_{\chi,f}}{T}\right) + \frac{\sigma}{2}Q_f E^{\mu} + \chi\frac{N_c Q_f}{4\pi^2}\mu_{\chi,f}B^{\mu} \tag{4.10}$$

In Fig. 4.10 we show such a comparison to demonstrate the influence of the potential relaxation effect on the CME current. The results suggest a very mild sensitivity, with the instant-CME scenario giving a stronger charge separation signal. The relaxation effect has two opposite impacts on the CME current: it delays the buildup of the CME current at the beginning, while on the other hand also delays the decay of the CME current (once it is there) with decreasing magnetic field. Due to the competition of the two, even this extreme scenario with 100%-relaxation only reduces the signal mildly. The realistic case should be somewhere in between these two curves.

It is worth emphasizing that in order to see the influence of potential relaxation effect, we start both scenarios, with or without relaxation effect, with exactly the same initial condition of axial and vector charge density, but no initial axial or vector currents, i.e., assuming no pre-thermal CME. One can expect that if with the pre-thermal CME current, the relaxation effect would slowdown the decay of the CME current, and help to develop more charge separation.

4.8 Finite Quark Mass and Strangeness Transport

Another question of both theoretical interest and experimental relevance, is whether
the strange quarks and antiquarks undergo the anomalous transport and contribute to
certain observables. For the u and d quarks in the QGP, their masses are negligibly
small compared with temperature scale, therefore they can be well approximated
as chiral fermions. This is not the case for the strange quark or antiquark, whose
mass is not small compared with temperature, $\frac{M_s}{T} \sim \frac{1}{3}$. A finite mass will lead to
random flipping of chirality and cause dissipation of nonzero axial charge. As a
result, the anomalous transport will be suppressed. It has been recently argued that
such finite mass effect scales as $\left(\frac{M_s}{T}\right)^2$ [28, 29], which in the case of strange quark
would not be a severe suppression. This would thus leave certain room for potential
contributions to anomalous transport processes from strangeness sector.

To get some quantitative insight on the impact of this issue, let us consider two
extreme cases: the case with strange quark experiencing the anomalous transport,
equally as the light quarks (aka. "3-flavor" case); or the case with no anomalous
transport at all for s quarks (aka. "2-flavor" case).

From Fig. 4.11 (left panel) one can find that the charge separation signal of
either pions only or all charged particles is insensitive to strangeness contributions, whereas the charge separation signal of kaons is extremely sensitive to
any strangeness contributions. On the right panel we show the ratio of $K^\pm$ H-
correlations to $\pi^\pm$ H-correlations: the reality could well lie in between these two
extreme scenarios, and a precise experimental measurement of this ratio will provide
highly valuable insight into the anomalous transport of the strangeness sector.

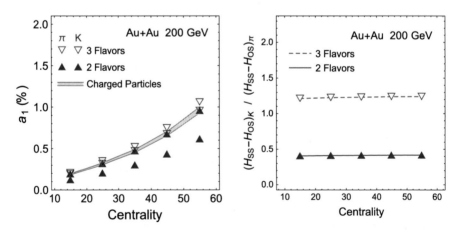

Fig. 4.11 CME signal of identified particles for the "2-flavor case" (upward solid triangles) versus
the "3-flavor case" (downward open triangles). (left) Charge separation a_1 of charged pions (red
symbols) and charged kaons (red symbols) for different centrality, with the gray band for the charge
separation a_1 of all charged particles. (right) The corresponding ratio of $K^\pm$ H-correlations to $\pi^\pm$
H-correlations versus centrality

4.9 Discussions on the Evolution of Pre-Hydro Charge Separation

Finally, we discuss the potential contribution from the pre-hydro charge separation. As discussed before, the magnetic field is very strong at the early time, even before the start time of hydrodynamic evolution. The CME is a general transport phenomenon that would occur both in the equilibrium and the out-of-equilibrium setting. It is therefore conceivable that there could already be CME-induced charge separation during the pre-hydro stage. Indeed there have been various studies of pre-equilibrium generation of charge separation, see e.g. [11, 30–35]. This implies that by the start of hydrodynamics, the density and current could already become nontrivial.

Such pre-hydro charge separation can be naturally integrated with AVFD framework as initial conditions for the corresponding fermion densities and currents at the hydro initial time τ_0. In the present section, we investigate how the nontrivial initial conditions via charge density dipole and charge current along **B** field from the pre-hydro evolution will propagate through the hydrodynamic stage toward final hadron observables. To do this, one can recast RH/LH currents into vector/axial currents and rewrite Eqs. (3.9)–(3.12) as

$$\hat{D}_\mu J_f^\mu = 0 \,, \tag{4.11}$$

$$J_f^\mu = n_f u^\mu + v_f^\mu \,, \tag{4.12}$$

$$\Delta^\mu_{\ \nu} \hat{d} \left(v_f^\nu \right) = -\frac{1}{\tau_r} \left[\left(v_f^\mu \right) - \left(v_f^\mu \right)_{NS} \right] \,, \tag{4.13}$$

$$\left(v_f^\mu \right)_{NS} = \sigma T \Delta^{\mu\nu} \partial_\nu \left(\frac{\mu_f}{T} \right) \,, \tag{4.14}$$

and similar equations can be obtained for J_5, n_5.

To quantify the pre-hydro currents and charge dipole, we rescale them by the initial entropy density s_0 at τ_0, with dimensionless factor λ. When solving the propagation of initial currents, we start with the initial condition that

$$v_f|_{\tau=\tau_0} = J_{\text{ini},f} = \lambda_{\text{cur},f} \, s_0 \, \hat{y}, \tag{4.15}$$

while for initial charge dipole,

$$n_{\text{ini},f} = (\lambda_{0,f} + \lambda_{\text{dip},f} \sin\phi) \, s_0. \tag{4.16}$$

Also, as these pre-hydro currents/dipoles are due to the Chiral Magnetic Effect, one could expect that they should be proportional to the corresponding quarks' electric charge:

$$\lambda_{\text{dip}} \equiv \lambda_{\text{dip},u} = -2\lambda_{\text{dip},d}, \tag{4.17}$$

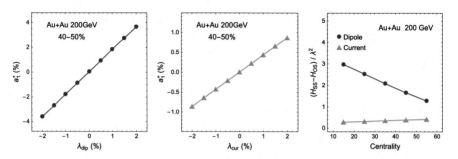

Fig. 4.12 Final charge separation signal computed via AVFD with nontrivial initial conditions from pre-hydro CME-induced charge dipole or current: (left) Charge separation a_1 caused by nonzero initial dipole; (middle) Charge separation a_1 caused by nonzero initial current; (right) Centrality dependence of H^{OS-SS} correlations due to nonzero initial dipoles (red) or currents (green), normalized by the respective initial condition parameters

$$\lambda_{\text{cur}} \equiv \lambda_{\text{cur},u} = -2\lambda_{\text{cur},d}. \tag{4.18}$$

Starting from such initial conditions with given λ_{dip} or λ_{cur}, we solve the AVFD equations and compute the final state hadron charge separation signal a_1. Figure 4.12 shows the results for a_1 due to *only* pre-hydro charge dipole (left panel) or current (middle panel) for given initial condition parameters for a particular centrality. One finds that the final signal responds linearly to the initial charge separation from pre-hydro CME. In the right panel we show the final correlations H^{OS-SS} scaled by λ_{dip} or λ_{cur} for the two types of initial conditions for a variety of centrality. The results suggest that the pre-hydro charge separation could reach a level at several percent of the initial entropy density, its effect on final signal would contribute a substantial fraction of the experimental data (around the magnitude of $\sim 10^{-4}$). With realistic pre-equilibrium models that could quantitatively compute the CME-induced early charge separation, one could then use those results as input for AVFD to predict final observables. In Chap. 6 we will employ the Wigner function framework to semi-quantitatively estimate the order of magnitude of the such pre-hydro charge separation.

4.10 Summary of Smooth-AVFD Results

In this chapter, we employ the Anomalous-Viscous Fluid Dynamics (AVFD) framework on top of event-averaged/smooth hydro background, and then performed detailed study to quantify the charge separation signal from the Chiral Magnetic Effect in heavy-ion collisions. With such tool, we quantitatively investigate the sensitivity of CME signal to a series of key parameters, including the time dependence of the magnetic field, the initial axial charge, the viscous transport coefficients as well as the resonance decay contributions. With realistic initial conditions and magnetic field lifetime, the predicted CME signal is quantitatively consistent with measured charge separation data in 200 GeV Au–Au collisions. We further predict the CME observables for the upcoming isobaric (Ru–Ru v.s. Zr–Zr) collision experiment that could provide the critical test for the presence of the CME.

The AVFD framework has further allowed us to investigate the influence on the CME signal by several theoretical uncertainties. We find that the possible thermal relaxation effect on the CME transport current has a mild impact on the final signal. The potential contribution to anomalous transport from the strangeness sector could have a substantial and observable consequence on the charged kaon correlation signals. In addition with the AVFD we quantify the final state charge separation arising entirely from nontrivial charge or current initial conditions due to the pre-hydro CME contributions.

References

1. J. Bloczynski, X.G. Huang, X. Zhang, J. Liao, Phys. Lett. **B718**, 1529 (2013). https://doi.org/10.1016/j.physletb.2012.12.030
2. L. McLerran, V. Skokov, Nucl. Phys. **A929**, 184 (2014). https://doi.org/10.1016/j.nuclphysa.2014.05.008
3. U. Gursoy, D. Kharzeev, K. Rajagopal, Phys. Rev. **C89**(5), 054905 (2014). https://doi.org/10.1103/PhysRevC.89.054905
4. K. Tuchin, Phys. Rev. **C93**(1), 014905 (2016). https://doi.org/10.1103/PhysRevC.93.014905
5. G. Inghirami, L. Del Zanna, A. Beraudo, M.H. Moghaddam, F. Becattini, M. Bleicher, Eur. Phys. J. **C76**(12), 659 (2016). https://doi.org/10.1140/epjc/s10052-016-4516-8
6. D. Kharzeev, A. Krasnitz, R. Venugopalan, Phys. Lett. **B545**, 298 (2002). https://doi.org/10.1016/S0370-2693(02)02630-8
7. B. Muller, A. Schafer, Phys. Rev. **C82**, 057902 (2010). https://doi.org/10.1103/PhysRevC.82.057902
8. Y. Hirono, T. Hirano, D.E. Kharzeev (2014). arXiv:1412.0311
9. A.H. Rezaeian, M. Siddikov, M. Van de Klundert, R. Venugopalan, Phys. Rev. **D87**(3), 034002 (2013). https://doi.org/10.1103/PhysRevD.87.034002
10. H. Kowalski, T. Lappi, R. Venugopalan, Phys. Rev. Lett. **100**, 022303 (2008). https://doi.org/10.1103/PhysRevLett.100.022303
11. M. Mace, S. Schlichting, R. Venugopalan, Phys. Rev. **D93**(7), 074036 (2016). https://doi.org/10.1103/PhysRevD.93.074036
12. G.S. Denicol, H. Niemi, I. Bouras, E. Molnar, Z. Xu, D.H. Rischke, C. Greiner, Phys. Rev. **D89**(7), 074005 (2014). https://doi.org/10.1103/PhysRevD.89.074005
13. A. Amato, G. Aarts, C. Allton, P. Giudice, S. Hands, J.I. Skullerud, Phys. Rev. Lett. **111**(17), 172001 (2013). https://doi.org/10.1103/PhysRevLett.111.172001
14. L. Adamczyk et al., Phys. Rev. Lett. **113**, 052302 (2014). https://doi.org/10.1103/PhysRevLett.113.052302
15. B.I. Abelev et al., Phys. Rev. Lett. **103**, 251601 (2009). https://doi.org/10.1103/PhysRevLett.103.251601
16. B.I. Abelev et al., Phys. Rev. **C81**, 054908 (2010). https://doi.org/10.1103/PhysRevC.81.054908
17. L. Adamczyk et al., Phys. Rev. **C88**(6), 064911 (2013). https://doi.org/10.1103/PhysRevC.88.064911
18. S.A. Voloshin, Phys. Rev. **C70**, 057901 (2004). https://doi.org/10.1103/PhysRevC.70.057901

19. V. Koch, S. Schlichting, V. Skokov, P. Sorensen, J. Thomas, S. Voloshin, G. Wang, H.U. Yee, Chin. Phys. **C41**(7), 072001 (2017). https://doi.org/10.1088/1674-1137/41/7/072001
20. W.T. Deng, X.G. Huang, G.L. Ma, G. Wang, Phys. Rev. **C94**, 041901 (2016). https://doi.org/10.1103/PhysRevC.94.041901
21. F. Wen, J. Bryon, L. Wen, G. Wang, Chin. Phys. **C42**(1), 014001 (2018). https://doi.org/10.1088/1674-1137/42/1/014001
22. N. Magdy, S. Shi, J. Liao, N. Ajitanand, R.A. Lacey, Phys. Rev. **C97**(6), 061901 (2018). https://doi.org/10.1103/PhysRevC.97.061901
23. N. Magdy, S. Shi, J. Liao, P. Liu, R.A. Lacey, Phys. Rev. **C98**(6), 061902 (2018). https://doi.org/10.1103/PhysRevC.98.061902
24. J. Zhao, H. Li, F. Wang, Eur. Phys. J. **C79**(2), 168 (2019). https://doi.org/10.1140/epjc/s10052-019-6671-1
25. H.J. Xu, J. Zhao, X. Wang, H. Li, Z.W. Lin, C. Shen, F. Wang, Chin. Phys. **C42**(8), 084103 (2018). https://doi.org/10.1088/1674-1137/42/8/084103
26. J. Bloczynski, X.G. Huang, X. Zhang, J. Liao, Nucl. Phys. **A939**, 85 (2015). https://doi.org/10.1016/j.nuclphysa.2015.03.012
27. D.E. Kharzeev, M.A. Stephanov, H.U. Yee, Phys. Rev. **D95**(5), 051901 (2017). https://doi.org/10.1103/PhysRevD.95.051901
28. E.D. Guo, S. Lin, Phys. Rev. **D93**(10), 105001 (2016). https://doi.org/10.1103/PhysRevD.93.105001
29. Z. Wang, X. Guo, S. Shi, P. Zhuang (2019). arXiv:1903.03461 [hep-ph]
30. M. Mace, N. Mueller, S. Schlichting, S. Sharma, Phys. Rev. **D95**(3), 036023 (2017). https://doi.org/10.1103/PhysRevD.95.036023
31. N. Müller, S. Schlichting, S. Sharma, Phys. Rev. Lett. **117**(14), 142301 (2016). https://doi.org/10.1103/PhysRevLett.117.142301
32. K. Fukushima, Phys. Rev. **D92**(5), 054009 (2015). https://doi.org/10.1103/PhysRevD.92.054009
33. Y. Sun, C.M. Ko, F. Li, Phys. Rev. **C94**(4), 045204 (2016). https://doi.org/10.1103/PhysRevC.94.045204
34. A. Huang, Y. Jiang, S. Shi, J. Liao, P. Zhuang, Phys. Lett. **B777**, 177 (2018). https://doi.org/10.1016/j.physletb.2017.12.025
35. G.L. Ma, B. Zhang, Phys. Lett. **B700**, 39 (2011). https://doi.org/10.1016/j.physletb.2011.04.057

Chapter 5
How Event-by-Event Fluctuations Influence CME Signal

In previous chapter, we estimated CME charge separation observables with smooth-AVFD framework, simulating chiral anomalous transportation based on event-averaged bulk background. However, flow observables are strongly influenced by event-by-event fluctuation of initial bulk profile—the fluctuating bulk geometry given non-vanishing odd-order azimuthal anisotropies, and would also have sizable influences on the even-order ones. For a precise quantitative study of CME signal, one should take such fluctuations into account, and we further develop the EbyE-AVFD framework, implementing anomalous transportation on top of the event-by-event hydrodynamics simulation provided by iEbE-VISHNU package. On the other hand, inherited from iEbE-VISHNU, the EbyE-AVFD simulates hadron cascade after freeze-out, including scattering and resonance decay. Decay of resonances naturally brings nontrivial contribution to two-particle correlations γ and δ, and is believed to be one of the most important sources on the non-CME background [1]. Thus, EbyE-AVFD framework would also have the advantage to partially include non-CME background.

In this chapter, we will discuss our quantitative study on CME signal based on fluctuating bulk background, especially on the upcoming isobaric (Ru–Ru vs. Zr–Zr) collisions. Before discussing the results, let us first mention some subtleties in event-by-event simulations one should pay attention to.

5.1 Subtleties Due to Event Plane De-correlation

In a collision event one can define three different coordinate frames: the Reaction Plane (RP), the Participant Plane (PP), and the Event Plane (EP). The Reaction Plane is determined by incoming nuclei, with its x-axis parallel to impact parameter b, the vector pointing from the center of one nucleus to that of the other. On the other hand, the Participant Plane is determined by the initial geometry of the hot

© Springer Nature Switzerland AG 2019
S. Shi, *Soft and Hard Probes of QCD Topological Structures in Relativistic Heavy-Ion Collisions*, Springer Theses,
https://doi.org/10.1007/978-3-030-25482-7_5

medium, which is strongly influenced by fluctuations due to the scattering process between incoming nucleons. The Participant Plane is defined in the way that its origin overlaps with the bulk center-of-mass, with the orientation maximizing the elliptic anisotropy ϵ_2 along y-axis. However, both RP and PP are simulation-based frames, in a realistic experimental event, one is not possible to know either the Reaction Plane or the Participant Plane. The orientational information of initial profile is estimated from azimuthal distribution of final state particles, with the Event Plane determined by maximizing elliptic flow v_2 along x-axis[1]:

$$\frac{\sum_k w_k(p_T)\, e^{2i\phi_k}}{\sum_k w_k(p_T)} \equiv v_2\, e^{2i\,\Psi_{\text{EP}}} , \tag{5.1}$$

where $w_k(p_T)$ is the p_T weights introduced aiming to improve correlations, and is typically taken as being proportional to p_T with some threshold, e.g., $p_T < 2$ GeV.

In experimental analysis, all orientation depending observables, including v_2 as well as γ-correlator, are measured with respect to Event Plane. However, the Event Plane would not perfectly reproduce RP or PP, due to bulk evolution or final state fluctuation, and more importantly finite number effect—in realistic experimental analysis, EP is constructed by finite number of reference particles ($N_{Ref} \lesssim 100$ in a typical 200 AGeV RHIC event), and there would be significant de-correlation between such planes.

5.1.1   Statistical Estimation of Event Plane Resolution

In order to get an intuitive idea, let us take a simplified statistical model to estimate the value of Event Plane Resolution. We consider N particles with azimuthal angle $\{\varphi_i\}$, with probability distribution satisfying $P(\varphi) = \frac{1}{2\pi}(1 + 2\,v_2\cos(2\varphi))$. Probability distribution of the Event Plane quantities $(C, S) \equiv (\sum_i \cos 2\phi_i/N, \sum_i \sin\phi_i/N)$ could be obtained by taking the approximation of central limit theorem:

$$P(C, S) = \frac{N}{\pi\sqrt{1 - 4v_2^2}} \exp\left[-\frac{(C - v_2)^2}{2((1/2)^2 - v_2^2)/N} - \frac{S^2}{2(1/2)^2/N} \right]. \tag{5.2}$$

Then with given C and S, observed elliptic flow V_2^{obs} and Event Plane Ψ_{EP} are determined via the relation $C \equiv V_2^{\text{obs}}\cos 2\Psi_{\text{EP}}$, $S \equiv V_2^{\text{obs}}\sin 2\Psi_{\text{EP}}$, and they follow the probability distribution:

---

[1] It might be worth noting that EP (PP) maximizing v_2 (ϵ_2) is also referred to as second-order Event (Participant) Plane. Similarly one can define third-order or even arbitrarily higher order planes.

Fig. 5.1 Statistic model estimated Event Plane de-correlation $\langle\cos(2\Psi_{EP})\rangle$ for given anisotropy v_2 and N_{Ref}. N_{Ref} represents the number of reference particles to determine Event Plane

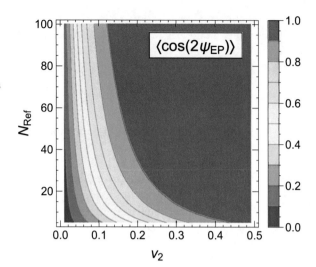

$$P(V_2^{obs}, \Psi_{EP}) = \frac{N V_2^{obs}}{\pi\sqrt{1-4v_2^2}} \exp\left[-\frac{(V_2^{obs}\cos 2\Psi_{EP}-v_2)^2}{(1-4v_2^2)/(2N)} - \frac{(V_2^{obs})^2 \sin^2 2\Psi_{EP}}{1/(2N)}\right].$$

(5.3)

Consequently, one can estimate the E.P. de-correlation via

$$\langle\cos(2\Psi_{EP})\rangle \equiv \int P(V_2^{obs}, \Psi_{EP}) \cos(2\Psi_{EP}) \, dV_2^{obs} d\Psi_{EP},$$

(5.4)

with the numerical value shown in Fig. 5.1. To maintain fair enough Event Plane Resolution, one would require a measurement with sufficiently large "true" anisotropy, as well as abundant reference particles to determine the Event Plane. It might be worth mentioning that the p_T weighting $w_k(p_T)$ would increase the mean v_2 of reference particles, and effectively improve the Event Plane Resolution.

5.1.2 How Event Plane De-correlation Influences CME Correlators

Now let us consider how CME signals would be influenced by Event Plane de-correlation. For simplification, we assume charged hadrons which have CME charge separation along Ψ_{CME}, while second-order Event Plane along $\Psi_{EP,2}$, follow the probability distribution:

$$f^\pm(p_T, \phi) = \frac{N(p_T)}{2\pi}[1 \pm 2a_{CME}(p_T)\cos(\phi - \Psi_{CME})$$

$$+ 2v_2(p_T)\cos(2\phi - 2\Psi_{EP,2}) + \cdots],$$

(5.5)

and we assume no "real" correlations between these particles, i.e., the joined probability distribution of two particles can be separated as the product of single particle distribution. In this simplified case, one can find that the orientation-dependent γ-correlator get reduced as

$$\gamma^{\alpha\beta} \equiv \langle \cos(\phi_\alpha + \phi_\beta - 2\Psi_{EP,2}) \rangle = a_{CME}^\alpha \cdot a_{CME}^\beta \cdot \cos(2\Psi_{CME} - 2\Psi_{EP,2}), \quad (5.6)$$

while the orientation-independent δ-correlator is not influenced by the de-correlation

$$\delta^{\alpha\beta} \equiv \langle \cos(\phi_\alpha - \phi_\beta) \rangle = a_{CME}^\alpha \cdot a_{CME}^\beta \cdot \qquad (5.7)$$

In the isobaric experiment comparing signals in two systems, the latter would reflect CME signal more ideally. Different from these correlators, the direct dipole is suppressed in a different way:

$$a_1^\pm \equiv \langle \sin(\phi - \Psi_{EP,2}) \rangle = \pm a_{CME} \cdot \cos(\Psi_{CME} - \Psi_{EP,2} - \pi/2), \qquad (5.8)$$

and one could find $\Delta\delta > (\langle a_1^\pm \rangle)^2 > \Delta\gamma$.

5.2 Simulation Setup

Similarly to smooth simulations, the Event-by-Event Anomalous-Viscous Fluid Dynamics (EbyE-AVFD) simulations need three components as the inputs/initial conditions: initial entropy density of the bulk background, space-time dependence of the magnetic field, and initial condition for axial charge density. We take fluctuating bulk initial entropy density generated Monte Carlo Glauber model, while axial charge density is assumed to be proportional to the entropy density. To extract the CME signal from the background, we take the simulation in four cases, differentiating in the amount of chirality imbalance that $n_5/s = 0, 0.05, 0.1, 0.2$, and measure γ and δ correlators for situations with purely background, or with different strength of CME signal.

In this study, EbyE-AVFD simulations are performed with two schemes of the magnetic field:

1. *smooth-B scheme* that magnetic field is homogeneous in spatial while decays in time, following

$$B(\tau, x) = \frac{B_0}{1 + \tau^2/\tau_B^2} \hat{y}, \qquad (5.9)$$

2. *fluctuating-B scheme* that while magnetic field decays in time in the same way as previous scheme, it has nontrivial spatial dependence:

$$B(\tau, x) = \frac{B(x)}{1 + \tau^2/\tau_B^2} , \tag{5.10}$$

taking lifetime $\tau_B = 0.6$ fm in both schemes and as in smooth-AVFD simulation. In latter scheme, the spatial profile is computed from proton distribution of the corresponding event:

$$B(x, y, 0) = \sum_{i \in \text{target}} B_t(x-x_i, y-y_i, 0-z_i) + \sum_{j \in \text{projectile}} B_p(x-x_j, y-y_j, 0-z_j), \tag{5.11}$$

where B_t (B_p) represents the magnetic field generated by a proton in the target (projectile), with center at (x_i, y_i, z_i) when $t = 0$. Also, we take into account the finite size of the proton, and assume its electric charge density as a Gaussian distribution with $\sigma = 0.5$ fm, corresponding to the root-mean-square radius of $\langle r^2 \rangle^{1/2} = 0.87$ fm.[2] With such setup, magnetic field generated by a given proton could be computed as

$$B_{t/p}(x, y, z) = (q \gamma) \int \frac{v_{t/p} \times (r - r')}{[(x - x')^2 + (y - y')^2 + \gamma^2(z - z')^2]^{3/2}} \rho(r'-r_i) d^3 r'. \tag{5.12}$$

Here we take the projectile velocity $v_p = +\sqrt{1 - \gamma^{-2}}\hat{z}$ along the positive z-direction, while the target velocity $v_t = -\sqrt{1 - \gamma^{-2}}\hat{z}$ along the negative z-direction.

Meanwhile, for comparison with experimental data, observables are measured for particles with kinematic region $0.15 < p_T < 3$ GeV, $|\eta| < 0.5$. While Event Planes are determined with reference particles in the same p_T-regime but another rapidity range $1.5 < |\eta_{\text{Ref}}| < 2.5$, via

$$q_2 e^{2i\Psi_{EP}} \equiv \left[\sum_{k \in \text{Ref}} w_k e^{2i\phi_k} \right] / \left[\sum_{k \in \text{Ref}} w_k \right], \tag{5.13}$$

with the p_T weighting $w_k \equiv p_T$, and, conventionally, the elliptic anisotropy for reference particles is denoted as q_2.

5.3 CME Signal in 200 GeV Au–Au Collisions

As the first step, we study how CME signal is influenced by event-by-event fluctuation in the well-studied Au–Au system. Let us examine the quantitative

---

[2]It has been tested that while the magnetic field for a particular point could be sensitive to the choice of σ, the averaged magnetic field, for instance, over the region $x^2 + y^2 < 9$ fm^2, is insensitive to σ.

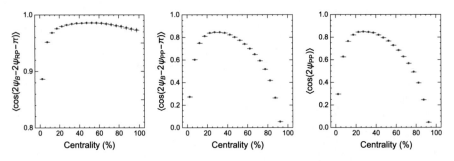

Fig. 5.2 Plane de-correlation factors for magnetic field direction Ψ_B versus Reaction Plane (left), Ψ_B versus Participant Plane (middle), and Participant Plane versus Reaction Plane (right) for event-by-event initial conditions. Centrality class is determined by total initial entropy, while magnetic field direction is determined from averaged B-field over the region $x^2 + y^2 < 9$ fm^2

properties of initial conditions, especially de-correlation between different directions. We sampled $\sim 10^7$ event-by-event initial configurations of entropy density and magnetic field, binned them into "centrality classes" via total entropy, and obtained de-correlation factors for different centrality bins. From Fig. 5.2 one can find that with $\langle \cos(2\Psi_B - \Psi_{RP} - \pi) \rangle$ approaching unity, and it is safe to take the magnetic field along y-axis in Reaction Plane. Meanwhile, the Participant Plane $\langle \cos(2\Psi_{PP} - \Psi_{RP}) \rangle$ remains in a very reasonable range $\gtrsim 0.7$ for the middle central collisions, one can expect good correlation between magnetic field direction and the bulk geometry. Hence, in the simulation, we focus on the 50–60% centrality class, which maximized the projected magnetic field. Also, given the stable B-field direction, it would be safe to run simulation with smooth-B scheme, by taking $eB_0 = 0.096$ GeV^2.

5.3.1 γ and δ Correlators

First of all, let us measure the γ and δ correlators for events in these different cases. While δ is plane-independent, we measure γ correlator by two methods, with respect to the Reaction Plane and to the Event Plane. As shown in Fig. 5.3, one can see clearly that $\gamma\{RP\}$ is greater than $\gamma\{EP\}$. Both δ and γ, with respect to both Reaction Plane and Event Plane, change proportionally to $(n_5/s)^2$, as expected by CME. Hence we fit the (n_5/s) dependence of correlation by the relation

$$y = y_{bkg} + y_{CME} \cdot (n_5/s)^2 , \tag{5.14}$$

where the quadratic coefficient y_{CME} reflects the CME charge separation.

It might be worth mentioning that it has been checked explicitly that the quadratic coefficients y_{CME} follow a simple proportional relation, with the factor close to the Event Plane Resolution.

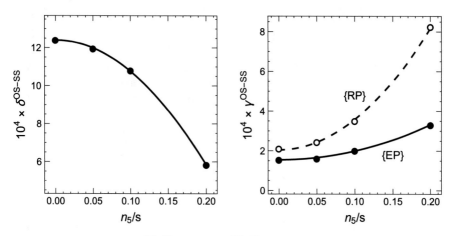

Fig. 5.3 CME correlators δ^{OS-SS} (left) and γ^{OS-SS} (right) for cases with different chirality imbalance n_5/s. γ-correlators represented by solid circles are measured with respect to Event Plane, while those represented by open circles are measured with respect to Reaction Plane. Fitting curves, with the form $y = y_{bkg} + y_{CME} \cdot (n_5/s)^2$, are also shown

$$\frac{y_{CME}[\gamma^{OS}\{RP\}]}{y_{CME}[\delta^{OS}]} = \frac{y_{CME}[\gamma^{SS}\{RP\}]}{y_{CME}[\delta^{SS}]} = \frac{y_{CME}[\gamma^{OS-SS}\{RP\}]}{y_{CME}[\delta^{OS-SS}]} = -0.93 \pm 0.02 \,,$$

$$\frac{y_{CME}[\gamma^{OS}\{EP\}]}{y_{CME}[\delta^{OS}]} = \frac{y_{CME}[\gamma^{SS}\{EP\}]}{y_{CME}[\delta^{SS}]} = \frac{y_{CME}[\gamma^{OS-SS}\{EP\}]}{y_{CME}[\delta^{OS-SS}]} = -0.27 \pm 0.02 \,.$$

5.3.2 Event-Shape Engineering

From the above subsection one can see in cases with different amount of chirality imbalance, both δ and γ correlators behave exactly as what is expected by CME. However, one key question in confirming CME in experiment data is to disentangle the signal from the background. Let us mention again that background analysis (see, e.g., [2, 3]) indicates that $y_{bkg} \propto v_2$. One natural way to tell signal from the background is to obtain γ for events with vanishing v_2, either by directly measuring γ for events in vanishing v_2 bin or by obtaining the intercept of v_2-γ correlation. This motivates the Event-Shape Engineering (ESE) analysis of the correlators.

However, as mentioned previously, the γ correlator would be suppressed by the de-correlation factor $\langle\cos(2\Psi_{CME} - 2\Psi_{EP,2} - \pi)\rangle$, while this de-correlation factor could also have nontrivial v_2 dependence. Thus, before doing the analysis, let us first examine the v_2 dependence of de-correlation factors.

In experimental analysis, there are two different methods to bin the events: (1) the q_2-*method* that events are classified into different q_2 bins, determined by reference particles, and obtain corresponding $v_2\{EP\}$ as well as any other observables for each bin; (2) the v_2-*method* that events are classified into different $v_2\{EP\}$ bins, determined by particles of interest, and obtain corresponding observables for each bin. Quantitative values for such de-correlation factors are shown in Fig. 5.4. One can see from the left panel that while binning with q_2, one would have vanishing resolution $\langle\cos(2\Psi_{EP,2})\rangle$, indicating totally random Event Plane angle and CME signal would also be vanishing. On the other hand, when employing the v_2-method, such resolution factor remains nonzero even for events with vanishing $v_2\{EP\}$. Consequently, compared to the q_2-method, v_2-method has the advantage of being possible to extract the CME signal from the intercept of v_2-γ correlation.

On the other hand, one would be interested in the possible v_2 dependence of the *pure* CME signal. In the left panel of Fig. 5.5 we show CME charge separation a_1 measured with respect to Reaction Plane. One can see the separation $a_1\{RP\}$ decreases with v_2, likely due to the geometrical effect that larger v_2 corresponds to longer length in y-direction, which makes it more difficult to develop sizable charge separation. Such behavior is also observed in the right panel showing

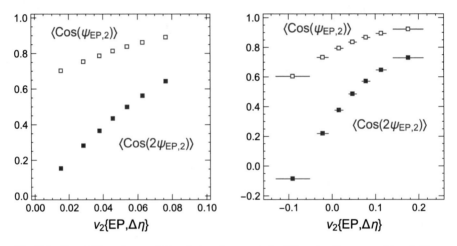

Fig. 5.4 Event Plane de-correlation factors for events in different reference q_2 bins (left) or different POI v_2 bins (right)

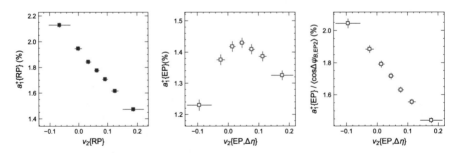

Fig. 5.5 CME charge separation a_1 measured with respect to Reaction Plane (left) and Event Plane (middle). Right panel shows $a_1\{EP\}$ corrected by resolution $\langle\cos(\Psi_{EP})\rangle$. All these three panels correspond to the case with strongest chirality imbalance $n_5/s = 0.2$

$a_1\{EP\}$ corrected by resolution $\langle\cos(\Psi_{EP})\rangle$, while $a_1\{EP\}$ itself (middle panel) is affected by both geometry and de-correlation, has weaker but more complicated v_2 dependence.

With all these preparation, we are now ready to examine the feasibility of using the intercept of $v_2 - \gamma$ correlation as CME signal. We measure γ correlator with respect to the Reaction Plane, where Event Plane de-correlation effect is avoided. In Fig. 5.6 one can see clearly the separation of intercepts for cases with different amount of chirality imbalance. Also, the intercept is consistent with zero while there is no CME, and grows quadratically with chirality imbalance. On the other hand, one can see the slope changes in different cases—which is expected to be constant—reflecting the v_2 dependence of a_1. Again, these are measurements with respect to the Reaction Plane, which is an over-idealized case. To be compared with experimental analysis, let us measure the $v_2 - \gamma$ correlation with respect to Event Plane.

Eventually, in Fig. 5.7 we show $v_2 - \gamma$ correlation with respect to Event Plane, as well as corresponding slope and intercept. Due to the Event Plane de-correlation effect, they are less separated compared to the idealized case. However, one can still see clearly that intercept fairly tells the CME signal from the background.

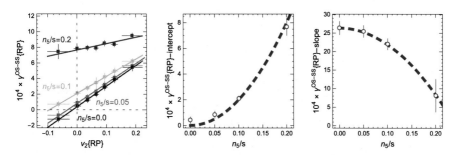

Fig. 5.6 (left) v_2-γ correlation, both measured with respect to Reaction Plane. Corresponding intercepts and slopes are shown in middle and right panels, respectively

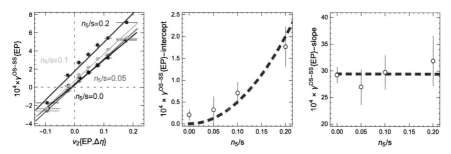

Fig. 5.7 (left) v_2-γ correlation, both measured with respect to Event Plane. Corresponding intercepts and slopes are shown in middle and right panels, respectively

To conclude we find that binning the events with $v_2\{EP\}$, elliptic flow for the particle of interest, instead of reference anisotropy q_2, could serve better the goal of disentangling CME signal from the background. We expect that this could be an efficient way to measure directly the CME signal.

5.4   CME in 200 GeV Isobaric Collisions

With the knowledge of Au–Au system, we move on to the quantitative study of fluctuation effects in isobaric ($^{96}_{44}$Ru–$^{96}_{44}$Ru versus $^{96}_{40}$Zr–$^{96}_{40}$Zr) collisions. As emphasized previously, the goal of IsoBar Program is to provide a crucial test of the CME signal. In these collisions, two contrast systems are expected to have similar non-CME background but different CME signal. By measuring the difference, one can disentangle the CME signal from the background.

It might be worth emphasizing that such expectation is based on the hypothesis that these two collisional systems have *identical* non-CME background. Thus, one key question is to quantify how identical the two systems are. This question has attracted intensive investigations (see, e.g., [4–6]), with the major uncertainty from the deformation of the nuclei. Taking the deformed Woods–Saxon distribution:

$$\rho_{\text{WS}}(r, \theta) \propto \frac{1}{1 + \exp[(r - R - \beta_2 Y_2^0(\theta)R)/a]} \, , \qquad (5.15)$$

with nucleus radii $R[\text{Ru}] = 5.085$ fm, $R[\text{Zr}] = 5.020$ fm, and skin depths $a[\text{Ru}] = a[\text{Zr}] = 0.46$ fm. There are two available sources of nuclei deformation [7]:

1. WS1 scheme: $e - A$ scattering experiments [8, 9] suggested that Ru is more deformed ($\beta_2[\text{Ru}] = 0.158$) than Zr ($\beta_2[\text{Zr}] = 0.08$);
2. WS2 scheme: a comprehensive model study [10] indicated $\beta_2[\text{Ru}] = 0.05$, $\beta_2[\text{Zr}] = 0.217$, where Zr is more deformed; for comparison, we set another scheme assuming no deformation, i.e., both of them are spherical;
3. WS0 scheme: $\beta_2[\text{Ru}] = 0$, $\beta_2[\text{Zr}] = 0$.

In Fig. 5.8 we show comparison of ellipticity and projected magnetic field between these two systems. In good agreement with aforementioned studies, we find notable differences of bulk geometry might occur in 0–20% most central collisions. However, we note that it shall be safe to treat the bulk background as identical for the centrality range of, e.g., 30–60%, which is more interested in CME studies due to their large projected magnetic field.

To further relieve concerns on potential differences in the bulk background, we'd propose a fully contrast method to analyze experimental data: In experiment analysis, one can classify the events into two dimensional bins spanned by particle multiplicity N_{ch} and elliptic flow v_2/q_2. With the premise of having sufficient statistics, the bins are required to select narrow enough range of N_{ch} as well as v_2/q_2, to ensure that events from Ru–Ru and Zr–Zr collisions should have the same

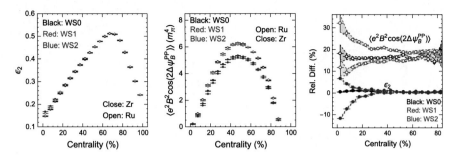

Fig. 5.8 Initial ellipticity ϵ_2 (left) and projected magnetic field (middle) along Participant Plane in IsoBar system. Right panel shows the relative difference $\frac{X[\mathrm{Ru}] - X[\mathrm{Zr}]}{(X[\mathrm{Ru}] - X[\mathrm{Zr}])/2}$ of ellipticity and magnetic field in these systems. See text for detail of three different deformation schemes

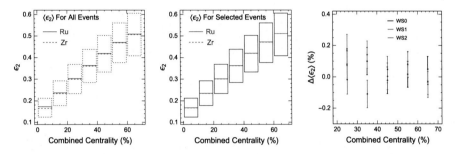

Fig. 5.9 Mean initial ellipticity $\langle \epsilon_2 \rangle$ for Ru (red solid) and Zr (blue dashed), by including (left) all events or (middle) only events in the selected ϵ_2 range, indicated by the black boxes. Relative difference of $\langle \epsilon_2 \rangle$ in these systems is shown in right panel

mean value of multiplicity and elliptic flow, hence are expected to be highly identical in the non-CME background. We expect that the comparison of γ and δ correlators for events in such bins would provide a highly decisive test of CME signal. Such idea is illustrated in Fig. 5.9: we classify minimal bias initial conditions into ten entropy (multiplicity) sets, roughly corresponding to ten centrality classes, and select events with ϵ_2 range around $(\langle \epsilon_2 \rangle_{\mathrm{Ru}} + \langle \epsilon_2 \rangle_{\mathrm{Zr}})/2$, with full width taken as one standard deviation of the distribution (covering $\sim$38% of total statistic). One can see that with such selection, even the most central events become much identical.

In our EbyE-AVFD simulations, we take the WS1-scheme which is estimated from experimental observables, and corresponds to the most conservative **B**-field difference. Also, in order to make more realistic quantification of CME signal, we employed the *fluctuating-B scheme* with nontrivial, fluctuating spatial dependence of **B**-field computed from protons' position. We focus on the events with $32 < N_{ch,|\eta|<0.5} < 48$, roughly corresponding to 40–50% centrality range, and make the selection of geometry by reference particle elliptic anisotropy q_2 (Fig. 5.10).

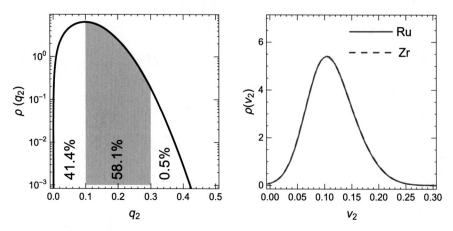

Fig. 5.10 (left) Statistic distribution of event-wise q_2, the gray band indicates event-selection range $0.1 < q_2 < 0.3$, corresponding to event class with 0.5–58.6% highest q_2. (right) Statistic distribution of event-wise elliptic flow v_2 of selected events for Ru–Ru (red solid) and Zr–Zr (blue dashed) collisions

5.4.1 q_2 Selection

As shown in Fig. 5.8, for the 40–50% centrality class, difference of initial ellipticity in the isobaric system is rather small, and we are safe to take a wider range $0.1 < q_2 < 0.3$. Such selection corresponds to events that have 0.5–58.6% highest q_2. In Table 5.1 one can find that with such selection we effectively improve the Event Plane Resolution. Again, we note that the statistical distribution of v_2 is identical between such system.

Comparing γ and δ correlators for events from the full q_2 range (Fig. 5.11) with selected range $0.1 < q_2 < 0.3$ (Fig. 5.12), we find such selection effectively improves the distinction of γ correlators, and expect it to be a useful technique in experimental analysis. On the other hand, one can also find δ correlator turns out to be a more ideal observable to distinguish CME signal in these systems.

Table 5.1 De-correlation factors for events with or without q_2 cut

q_2 range	Full range	$0.1 < q_2 < 0.3$
$\langle\cos(2\Psi_{EP} - 2\Psi_{RP})\rangle$	0.37	0.48
$\langle\cos(2\Psi_{PP} - 2\Psi_{RP})\rangle$	0.66	0.69
$\langle\cos(2\Psi_{EP} - 2\Psi_{PP})\rangle$	0.57	0.70
$\langle\cos(2\Psi_{B} - 2\Psi_{RP} - \pi)\rangle$	0.98	0.98
$\langle\cos(2\Psi_{B} - 2\Psi_{PP} - \pi)\rangle$	0.66	0.68
$\langle\cos(2\Psi_{B} - 2\Psi_{EP} - \pi)\rangle$	0.37	0.48

Ψ_B is determined by the magnetic field averaged over the region $x^2 + y^2 < 9\,\text{fm}^2$

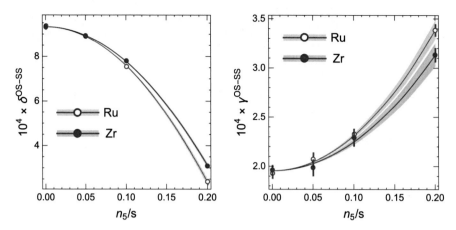

Fig. 5.11 CME correlators δ (left) and γ (right) for Ru–Ru (red open circle) and Zr–Zr (blue solid circle) collisions. Curves represent fitting result, while the bands include uncertainties of quadratic parameters

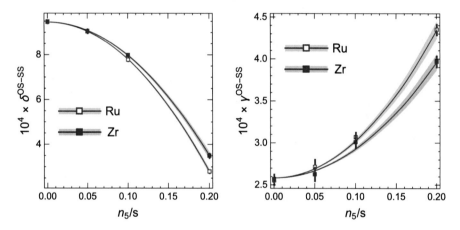

Fig. 5.12 CME correlators δ (left) and γ (right) for Ru–Ru (red open circle) and Zr–Zr (blue solid circle) collisions. Curves represent fitting result, while the bands include uncertainties of quadratic parameters. Only events within the range $0.1 < q_2 < 0.3$ are selected

5.5 Summary of the Event-by-Event AVFD Results

We summarize this chapter by emphasizing again on the key messages. According to the analysis of CME signal in 50–60% 200 GeV Au–Au collisions, we find that binning the events with $v_2\{EP\}$, elliptic flow for the particle of interest, instead of reference anisotropy q_2, could serve better the goal of disentangling CME signal from the background. Additionally, our study on CME observables in isobaric (Ru–Ru vs. Zr–Zr) collisions suggests that the comparison between observables within sufficiently narrow $N_{ch} \otimes q_2$ bins would provide a highly decisive test of CME signal. Meanwhile, being unaffected by the Event Plane de-correlation, we expect the δ correlator would be a more ideal observable to distinguish CME signal in these systems.

References

1. F. Wang, J. Zhao, Phys. Rev. **C95**(5), 051901 (2017). https://doi.org/10.1103/PhysRevC.95.051901
2. A. Bzdak, V. Koch, J. Liao, Phys. Rev. **C83**, 014905 (2011). https://doi.org/10.1103/PhysRevC.83.014905
3. A. Bzdak, V. Koch, J. Liao, Lect. Notes Phys. **871**, 503 (2013). https://doi.org/10.1007/978-3-642-37305-3_19
4. W.T. Deng, X.G. Huang, G.L. Ma, G. Wang, Phys. Rev. **C94**, 041901 (2016). https://doi.org/10.1103/PhysRevC.94.041901
5. W.T. Deng, X.G. Huang, G.L. Ma, G. Wang, Phys. Rev. **C97**(4), 044901 (2018). https://doi.org/10.1103/PhysRevC.97.044901
6. H.J. Xu, X. Wang, H. Li, J. Zhao, Z.W. Lin, C. Shen, F. Wang, Phys. Rev. Lett. **121**(2), 022301 (2018). https://doi.org/10.1103/PhysRevLett.121.022301
7. Q.Y. Shou, Y.G. Ma, P. Sorensen, A.H. Tang, F. Videbæk, H. Wang, Phys. Lett. **B749**, 215 (2015). https://doi.org/10.1016/j.physletb.2015.07.078
8. S. Raman, C.W.G. Nestor Jr., P. Tikkanen, At. Data Nucl. Data Tables **78**, 1 (2001). https://doi.org/10.1006/adnd.2001.0858
9. B. Pritychenko, M. Birch, B. Singh, M. Horoi, At. Data Nucl. Data Tables **107**, 1 (2016). https://doi.org/10.1016/j.adt.2015.10.001, https://doi.org/10.1016/j.adt.2016.08.002. [Erratum: At. Data Nucl. Data Tables 114,371(2017)]
10. P. Moller, J.R. Nix, W.D. Myers, W.J. Swiatecki, At. Data Nucl. Data Tables **59**, 185 (1995). https://doi.org/10.1006/adnd.1995.1002

Chapter 6
The Chiral Magnetic Effect in Pre-equilibrium Stage

As emphasized in Sect. 4.9, one important input of the AVFD simulation is the pre-equilibrium CME current/dipole. Given the short expected lifetime of the magnetic field, the CME current and charge dipole may be an important source of final observed CME signal. To study the CME current/dipole induced in pre-hydro stage, one requires an anomalous chiral transport theory to describe the out-of-equilibrium situation. The natural framework is the kinetic theory based on transport equations for the phase space distribution function of such a system. Different from usual classical kinetic theory, a proper description of the chiral fermions must account for intrinsic quantum and relativistic effects. A lot of progress has been achieved lately to develop such a chiral kinetic theory, see e.g. [1–12]. There also exist a lot of phenomenological interests and attempts to study anomalous chiral transport in the out-of-equilibrium setting [13–18]. The transport theory of chiral fermions, however, bears unusual subtlety and poses a number of challenges, particularly related to Lorentz invariance and frame dependence. A resolution was developed in the 3D formulation of chiral kinetic theory [5, 6, 10], but the origin of such issues remains cloudy. It is highly desirable to develop a transport theory of chiral fermions in a completely covariant fashion and to identify the precise reason of these complications.

A natural approach is to derive the quantum transport equation for chiral fermions in the well-known Wigner function formalism by a systematic semi-classical expansion in terms of $\hbar$ [19–24]. Before we proceed, it should be emphasized that the present study is drastically different from the studies in [7–9] which focus on behaviors from specific near-equilibrium solutions to the Wigner function equations under constant background electromagnetic fields. Our goal is to develop a general covariant quantum transport equation at order $\mathbf{O}(\hbar)$ for a system of chiral fermions.

In this chapter, we systematically study the chiral kinetic theory of the chiral fermions system under external electromagnetic fields. We employ semi-classical expansion of the operators and functions, to derive the covariant chiral kinetic equation, and clarify the subtle and confusing frame dependence issues associated

© Springer Nature Switzerland AG 2019
S. Shi, *Soft and Hard Probes of QCD Topological Structures in Relativistic Heavy-Ion Collisions*, Springer Theses, https://doi.org/10.1007/978-3-030-25482-7_6

with the $\mathbf{O}(\hbar)$ correction. We further estimate the CME current/dipole induced in the pre-equilibrium stage as well as the final charge separation signal it causes.

6.1 Wigner Function Formalism

The Wigner function formalism [19, 25] is a gauge-invariant and Lorentz covariant theory to describe a Dirac field ψ with charge Q under the presence of a gauge field $A^\mu(x)$. The general gauge-invariant Wigner operator is defined as

$$\hat{W}_{\alpha\beta}(x, p) = \int \frac{d^4 y}{(2\pi)^4} e^{-\frac{i}{\hbar} p \cdot y} \overline{\psi}_\beta(x_+) U(x_+, x_-) \psi_\alpha(x_-), \tag{6.1}$$

where α and β are spinor indices. The gauge link U between $x_\pm = x \pm y/2$ is introduced to ensure the gauge invariance of the Wigner operator. It is defined as

$$U(x_+, x_-) = \mathcal{P} e^{-\frac{iQ}{\hbar} y^\mu \int_0^1 ds A_\mu(x - \frac{y}{2} + sy)}, \tag{6.2}$$

where the path-ordering operator $\mathcal{P}$ becomes trivial for Abelian A^μ fields. In this work, we keep the reduced Planck constant $\hbar$ to show quantum effect explicitly. Then one can construct the Winger function, as the expectation of the Wigner operator

$$W_{\alpha\beta}(x, p) = \left\langle \hat{W}_{\alpha\beta}(x, p) \right\rangle, \tag{6.3}$$

where $\langle \cdot \rangle$ means the expectation over an arbitrary quantum state, or even the average over an ensemble of quantum states.

In this work, we consider a collisionless system in a background electromagnetic field A^μ. In this case, the Wigner function satisfies the quantum kinetic equation [19]

$$\left(\mathbf{K} - m \right) W(x, p) = 0. \tag{6.4}$$

where $\mathbf{K} = \gamma^\mu \mathbf{K}_\mu$, $\mathbf{K}_\mu = \pi_\mu + \frac{1}{2} i \hbar \nabla_\mu$, and

$$\pi^\mu = p^\mu - \frac{1}{2} Q\hbar j_1 \left(\frac{1}{2} \hbar \triangle \right) F^{\mu\nu} \partial_\nu^p, \tag{6.5}$$

$$\nabla^\mu = \partial^\mu - Q j_0 \left(\frac{1}{2} \hbar \triangle \right) F^{\mu\nu} \partial_\nu^p. \tag{6.6}$$

Note that in the triangle operator $\triangle = \partial_x \cdot \partial_p$, ∂_x acts only on electromagnetic tensor $F^{\mu\nu} = \partial^\mu A^\nu - \partial^\nu A^\mu$, while ∂_p acts only on $W(x, p)$. In addition,

$j_0(x) = x^{-1}\sin(x)$ and $j_1(x) = x^{-2}\sin(x) - x^{-1}\cos(x)$ are the spherical Bessel functions which are generated by the y-integrations. In general combining with the Maxwell equation, the quantum kinetic equation of Wigner function Eq. (6.4) is equivalent to the full QED.

In order to connect Eq. (6.4) with classical kinetic theory, one needs to obtain explicitly the equations of all elements of the Wigner function, which is a 4×4 matrix. To this aim, we expand $W(x, p)$ in terms of the 16 independent generators of the Clifford algebra, choosing the convention basis as follows:

$$\Gamma^a = I, \ \gamma^\mu, \ i\gamma^5, \ \gamma^\mu\gamma^5, \ \sigma^{\mu\nu},$$

$$\Gamma_a = I, \ \gamma_\mu, \ -i\gamma_5, \ \gamma_5\gamma_\mu, \ \sigma_{\mu\nu}. \tag{6.7}$$

In this basis, the Wigner function is expanded as

$$W = \frac{1}{4}\left(\mathscr{F} + i\gamma^5\mathscr{P} + \gamma^\mu\mathscr{V}_\mu + \gamma^\mu\gamma^5\mathscr{A}_\mu + \frac{1}{2}\sigma^{\mu\nu}\mathscr{L}_{\mu\nu}\right), \tag{6.8}$$

6.2 Chiral Kinetic Equation

Substituting the decomposition form of the Wigner function (6.8) into the equation of motion (6.4), and taking the chiral limit $m = 0$, we obtained the quantum kinetic equations which are partially decoupled into two sets: a set of equations describing the evolution of scalar $\mathscr{F}$, pseudoscalar $\mathscr{P}$, and antisymmetry tensor $\mathscr{L}^{\mu\nu}$,

$$\pi_\mu\mathscr{F} + \frac{1}{2}\hbar\nabla^\nu\mathscr{L}_{\mu\nu} = 0,$$

$$\frac{1}{2}\hbar\nabla_\mu\mathscr{F} - \pi^\nu\mathscr{L}_{\mu\nu} = 0,$$

$$-\hbar\nabla_\mu\mathscr{P} + \epsilon_{\mu\nu\rho\sigma}\pi^\nu\mathscr{L}^{\rho\sigma} = 0, \tag{6.9}$$

$$\pi_\mu\mathscr{P} + \frac{1}{4}\hbar\epsilon_{\mu\nu\rho\sigma}\nabla^\nu\mathscr{L}^{\rho\sigma} = 0,$$

and another set for vector $\mathscr{V}_\mu$ and axial vector $\mathscr{A}_\mu$,

$$\pi^\mu\mathscr{V}_\mu = 0, \qquad \pi^\mu\mathscr{A}_\mu = 0,$$

$$\hbar\nabla^\mu\mathscr{V}_\mu = 0, \qquad \hbar\nabla^\mu\mathscr{A}_\mu = 0,$$

$$\hbar\epsilon_{\mu\nu\rho\sigma}\nabla^\rho\mathscr{V}^\sigma = 2(\pi_\mu\mathscr{A}_\nu - \pi_\nu\mathscr{A}_\mu), \tag{6.10}$$

$$\hbar\epsilon_{\mu\nu\rho\sigma}\nabla^\rho\mathscr{A}^\sigma = 2(\pi_\mu\mathscr{V}_\nu - \pi_\nu\mathscr{V}_\mu).$$

Serving the purpose of obtaining the transport theory for both normal and anomalous currents, we will focus on the evolution equations of chiral vector Eq. (6.10) in the following parts of this article. Noticing the symmetry between vector and axial vector terms, one could further simplify the above equations by introducing the "chiral basis" [9, 23] via

$$\mathscr{J}_\chi^\mu = \frac{1}{2}(\mathscr{V}^\mu - \chi \mathscr{A}^\mu),\qquad(6.11)$$

where $\chi = \pm 1$, corresponds to the chirality of massless fermion. Similar to $\mathscr{V}^\mu$ and $\mathscr{A}^\mu$, which are related to vector and axial currents, the chiral vectors $\mathscr{J}_\chi^\mu$ have the physical meaning corresponding to chiral current

$$j_\chi^\mu = \langle \bar{\psi}_\chi \gamma^\mu \gamma^5 \psi_\chi \rangle = \int d^4 p\, \mathscr{J}_\chi^\mu = \frac{1}{2}\left(j^\mu + \chi j_5^\mu\right).\qquad(6.12)$$

Here $\psi_\chi = P_\chi \psi$, $\bar{\psi}_\chi = \bar{\psi} P_{-\chi}$, $P_\chi = (1 + \chi \gamma^5)/2$ denote the chiral state and chiral projection operator, respectively. In such chiral basis, Eq. (6.10) can be further decomposed, in which the right-handed (RH) and left-handed (LH) components get decoupled:

$$\hbar \epsilon_{\mu\nu\rho\sigma} \nabla^\rho \mathscr{J}_\chi^\sigma = -2\chi (\pi_\mu \mathscr{J}_\nu^\chi - \pi_\nu \mathscr{J}_\mu^\chi),\qquad(6.13)$$

$$\pi^\mu \mathscr{J}_\mu^\chi = 0,\qquad(6.14)$$

$$\nabla^\mu \mathscr{J}_\mu^\chi = 0.\qquad(6.15)$$

The decoupling of the RH and LH components in these equations reflects a fundamental property of massless fermion systems: as can be seen in massless Dirac equations, the RH and LH fermions are not entangled, and one can treat them separately.

6.2.1   Covariant Chiral Kinetic Equation

To obtain the chiral kinetic equation (CKE), we first construct $\mathscr{J}_\chi^\mu$ which satisfies Eqs. (6.13)–(6.14) up to $\hbar^1$ order,

$$\mathscr{J}_\chi^\mu = p^\mu f_\chi \delta(p^2) + \hbar \chi Q \tilde{F}^{\mu\nu} p_\nu f_\chi^{(0)} \delta'(p^2) - \hbar \frac{\chi}{2p\cdot n}\epsilon^{\mu\nu\lambda\rho} n_\nu p_\lambda \left(\nabla_\rho f_\chi^{(0)}\right)\delta(p^2).$$

$$(6.16)$$

Here $\tilde{F}^{\mu\nu} = \frac{1}{2}\epsilon_{\mu\nu\rho\sigma} F^{\rho\sigma}$ is the dual tensor of $F^{\mu\nu}$, while $f_\chi = f_\chi^{(0)} + \hbar f_\chi^{(1)}$, which can also be decomposed into positive/negative energy parts, $f_\chi(x, p) = \sum_{\epsilon=\pm1} \theta(\epsilon p^0) f_\chi^\epsilon(x, \epsilon p)$. Mathematically, the auxiliary field n^μ can be an arbitrary time-like vector field with nontrivial space-time dependence, which corresponds to the frame dependence of CKE (details can be found in Ref. [26]). However, to avoid subtleties of comparing variables (via derivative) in a frame with nontrivial metric, let us consider a simplified case and take n^μ as a constant-homogeneous 4-vector u^μ. In this case the time and space derivatives of n^μ are trivial, and the CKE can be derived from Eq. (6.15) can be written as

$$0 = \delta \left(p^2 - \hbar \frac{\chi Q}{p \cdot u} (B \cdot p) \right) \cdot$$

$$\left\{ p^\rho \nabla_\rho - \hbar \frac{\chi Q}{2 (p \cdot u)^2} \epsilon^{\mu\nu\lambda\rho} E_\mu u_\nu p_\lambda \nabla_\rho + \hbar \frac{\chi Q}{2 p \cdot u} p_\lambda \left(\partial_\rho B^\lambda \right) \partial_p^\rho \right\} f_\chi,$$

(6.17)

where $E^\mu = F^{\mu\nu} u_\nu$, $B^\mu = \widetilde{F}^{\mu\nu} u_\nu$.

One could even simplify the case by choosing $n^\mu = u^\mu = (1, 0, 0, 0)$, which can be achieved by a proper Lorentz transformation. Besides, we note that here χ denotes the chiral instead of the helicity and f_χ^ϵ indicates the distribution function of a given chiral particle or antiparticle. We express the CKE in the helicity $h \equiv \epsilon \chi$ basis:

$$\left\{ \partial_t + \frac{1}{\sqrt{G}} \left(\widetilde{v} + \hbar \epsilon Q (\widetilde{v} \cdot \mathbf{b}_h) \mathbf{B} + \hbar \epsilon Q \widetilde{\mathbf{E}} \times \mathbf{b}_h \right) \cdot \nabla_{\mathbf{x}} \right.$$

$$\left. + \frac{\epsilon Q}{\sqrt{G}} \left(\widetilde{\mathbf{E}} + \widetilde{v} \times \mathbf{B} + \hbar \epsilon Q (\widetilde{\mathbf{E}} \cdot \mathbf{B}) \mathbf{b}_h \right) \cdot \nabla_{\mathbf{p}} \right\} f_h^\epsilon (x, \mathbf{p}) = 0.$$

(6.18)

with corresponding Jacobian, energy, group velocity as

$$\mathbf{b}_h = \epsilon \mathbf{b}_\chi = \epsilon \chi \frac{\mathbf{p}}{2|\mathbf{p}|^3}, \quad \sqrt{G} = (1 + \hbar \epsilon Q \mathbf{b}_h \cdot \mathbf{B}), \quad \widetilde{\mathbf{E}} = \mathbf{E} - \frac{1}{\epsilon Q} \nabla_{\mathbf{x}} E_{\mathbf{p}},$$

$$E_{\mathbf{p}} = |\mathbf{p}| (1 - \hbar \epsilon Q \mathbf{B} \cdot \mathbf{b}_h), \quad \widetilde{v} = \frac{\partial E_{\mathbf{p}}}{\partial \mathbf{p}} = \widehat{\mathbf{p}} (1 + 2 \hbar \epsilon Q \mathbf{B} \cdot \mathbf{b}_h) - \hbar \epsilon Q b_h \mathbf{B}.$$

(6.19)

We note that the Eq. (6.18) precisely reproduce the 3D chiral kinetic theory developed in [2–4, 10] and widely discussed in the literature. As already shown in many previous studies, this 3D version correctly gives the chiral anomaly relation, the Chiral Magnetic Effect, as well as other interesting chiral transport effects. Therefore, the more general 4D chiral transport equation also automatically includes all such phenomena as its natural consequences.

6.3 Modeling Out-of-Equilibrium CME

With the above obtained chiral kinetic equations, we now apply them for estimating the chiral magnetic current that can be generated during the early moments in heavy-ion collisions when the created dense partonic matter is still out-of-equilibrium while the magnetic field is the strongest. In passing, we note that there has been study of pre-thermal chiral magnetic effect using classical-statistical field simulations [13, 14, 16]. We also note that there has been attempt of applying chiral kinetic transport for describing long time evolution of the fireball assuming very long **B** field duration [17].

The partonic system at early time is characterized by the so-called saturation scale Q_s, on the order of 1–3 GeV for RHIC and the LHC [27]. We take $Q_s \simeq$ 2 GeV. According to recent kinetic studies of pre-equilibrium evolution (see e.g. [28–30]), while the system is initially gluon-dominated, the quarks are generated quickly on a time scale $\tau \sim 1/Q_s$ and then evolve toward thermal equilibrium. We will use a formation time $\tau_{in} = 0.1$ fm/c $\sim 1/Q_s$ as the starting time of our kinetic evolution of quark distributions until an end time of $\tau_f = 0.6$ fm/c $\sim 6/Q_s$ which is on the order of onset time for hydrodynamic evolution in heavy-ion collisions. We use the following quark initial distributions at $\tau = \tau_{in}$:

$$ f_{i0} = \lambda_i f_0 , \quad f_0 = n_0 \mathcal{F} (|\mathbf{p}|) \exp \left[-\frac{x^2}{R_x^2} - \frac{y^2}{R_y^2} - \frac{z^2}{R_z^2} \right] \qquad (6.20) $$

The spatial distribution is Gaussian, with three width parameters. The longitudinal width is set as $R_z = 1/(2Q_s)$. The transverse widths are determined by nuclear geometry, e.g. for AuAu collisions (with nuclear radius $R_A \simeq 6.3$ fm) at impact parameter b, $R_x \to (R_A - b/2)$ and $R_y \to \sqrt{R_A^2 - (b/2)^2}$. The initial momentum distribution $\mathcal{F}$ is not precisely known, so we will test the following three different forms and compare the results:

1. a Fermi-Dirac like form (FD)
   $\mathcal{F}_{FD} = \frac{1}{e^{(p-Q_s)/\Delta}+1}$ with $\Delta = 0.2$ GeV;
2. a soft-dominated Gaussian form (SG)
   $\mathcal{F}_{SG} = e^{-\frac{p^2}{Q_s^2}}$;
3. a hard-dominated Gaussian form (HG)
   $\mathcal{F}_{HG} = (\frac{cp}{Q_s})^2 e^{-(\frac{cp}{Q_s})^2}$ with $c = 1.65$.

The overall magnitude parameter n_0 is fixed by normalizing the quark number density at the fireball center via $\int_p f_0(\mathbf{p}, x = y = z = 0) \to \xi Q_s^3$ (see e.g. [30]) and we will vary the parameter ξ in a reasonable range to be consistent with that of typical initial condition used for hydrodynamic simulations. Finally the constant λ_i is used to specify the density difference across various species, namely u, $\bar{u}$, d, and $\bar{d}$ quarks with positive/negative helicity, respectively. We use the following choices: $\lambda_{u,+} = \lambda_{\bar{u},+} = \lambda_{d,+} = \lambda_{\bar{d},+} = 1 + \lambda_5$ while $\lambda_{u,-} = \lambda_{\bar{u},-} = \lambda_{d,-} = \lambda_{\bar{d},-} = 1 - \lambda_5$, where λ_5 controls the initial imbalance of opposite helicity fermions. We will vary λ_5 to examine the dependence of pre-thermal CME effect on such initial imbalance. In a realistic heavy-ion collision, the axial charge imbalance would be spatially fluctuating. Here the simple uniform imbalance is used to get a reasonable idea of how large the pre-thermal charge separation could be. With the presence of such imbalance, the relaxation time scale could become slightly different for fermions with opposite chirality. In this work we will use the same relaxation time scale for simplicity. Note also that the electric charge should be $q_u = -q_{\bar{u}} = \frac{2e}{3}$ and $q_d = -q_{\bar{d}} = -\frac{e}{3}$. For the rest of the section we focus on the case of impact parameter $b = 7.5$ fm corresponding roughly to 20–30% centrality class.

The time evolution of magnetic field $\mathbf{B} = B(\tau)\hat{\mathbf{y}}$ is not precisely determined. We will take an open attitude and compare a variety of possibilities proposed in the literature to provide a clear idea of the associated uncertainty. These include:

(a) The vacuum case $B(\tau) = \frac{B_0}{[1+(\tau/\bar{\tau})^2]^{3/2}}$ with $\bar{\tau} \simeq 0.076$ fm/c (referred to as VC hereafter) [31, 32];

(b) The medium-modified case assuming a conductivity value equal to lattice computed thermal value at $T \simeq 1.45T_c$ (referred to as MS-1 hereafter) [32];

(c) The medium-modified case assuming a conductivity value equal to 100 times the above-mentioned lattice value (referred to as MS-100 hereafter) [32];

(d) A widely used inverse-time-square parameterization $B(\tau) = \frac{B_0}{1+(\tau/\tau_B)^2}$ with $\tau_B = 0.1$ fm/c or $\tau_B = 0.6$ fm/c (referred to as SQ-01 and SQ-06 hereafter) [33–35];

(e) The dynamically evolving magnetic field $B(\tau)$ from the most recent magneto-hydrodynamic computation based on ECHO-QGP code (referred to as ECHO hereafter) [36].

The peak value of $eB_0 = 6m_\pi^2$ at $\tau = 0$ is set to be the same for all the above cases, taken from [37]. For a clear comparison, we show various $B(\tau)$ in Fig. 6.1. Note that the magnetic field in general is not homogenous in space. Nevertheless from the past event-by-event simulations (see e.g. [37]) such inhomogeneity of B field in most part of the overlapping zone in a heavy-ion collision is quite mild (except near the edge). We will assume a uniform magnetic field. This approximation shall be adequate to gain a semi-quantitative estimation of the magnitude of pre-thermal CME.

Let us now demonstrate the out-of-equilibrium charge separation due to CME by examining the transverse component $\mathbf{J}_\perp^Q$ of the electric charge current and the net charge density distribution n^Q on the x–y plane from the chiral kinetic transport solutions. To obtain concrete results, we've employed finite-difference numerical methods to explicitly evolve the kinetic equation in time. We use a spatial volume of $20 \times 20 \times 20$ fm^3 with grid size $\Delta r =$ fm and a finite time step $\Delta t = 0.001$ fm/c. In Fig. 6.2 we show the $\mathbf{J}_\perp^Q$ with the arrow indicating the direction of the current: it is evident that the current is aligned with magnetic field (along y-axis) and the magnitude is bigger in the area with larger local quark density. This CME-induced current will transport positive/negative charges in opposite direction and thus accumulate with time the separation of charges above/below the reaction plane. In Fig. 6.3 we show the net charge density distribution on the transverse plan at several time moments: indeed one clearly sees the gradual buildup of excessive positive charges on one side of the plane while negative charges on the other side, implying a growing charge separation with time.

Fig. 6.1 Comparison of various time-dependent magnetic field $B(\tau)$ normalized by $B_0 = B(\tau = 0)$

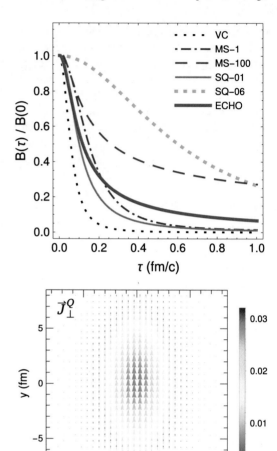

Fig. 6.2 Transverse charge current $\mathbf{J}_\perp^Q$ on the x–y plane at time $\tau = 0.2\,\text{fm/c}$ (computed with FD initial distribution, ECHO magnetic field, $\tau_R = 0.1\,\text{fm/c}$ and $\lambda_5 = 0.2$)

Let us then quantify the out-of-equilibrium charge separation effect and study its dependence on various ingredients in the modeling. To quantify this effect, we introduce a quantity $R_Q = N_Q/N_{total}$ defined as a ratio of the total number of net charge above reaction plane N_Q (with equal number but opposite net charge $-N_Q$ below reaction plane) to the total number of quarks and antiquarks N_{total} in the system. This ratio R_Q is shown in Fig. 6.4 as a function of time τ. The R_Q monotonically grows with time in all cases, reflecting the accumulation of charge separation from continuous CME transport. One also finds a strong dependence of this effect on the magnetic field evolution with time. For example, the ECHO magnetic field (from magnetohydrodynamic simulations) could produce a charge separation about *an order of magnitude larger* than the vacuum case. It is therefore

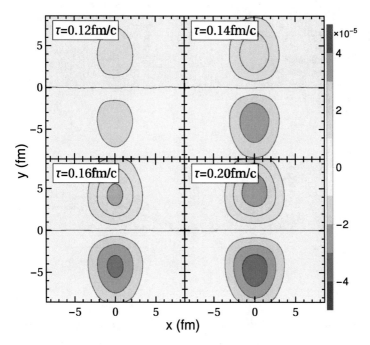

Fig. 6.3 Net charge density n^Q (normalized by ξQ_s^3) on the x–y plane at different time (computed with FD initial distribution, ECHO magnetic field, $\tau_R = 0.1$ fm/c and $\lambda_5 = 0.2$)

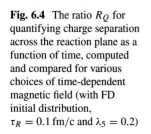

Fig. 6.4 The ratio R_Q for quantifying charge separation across the reaction plane as a function of time, computed and compared for various choices of time-dependent magnetic field (with FD initial distribution, $\tau_R = 0.1$ fm/c and $\lambda_5 = 0.2$)

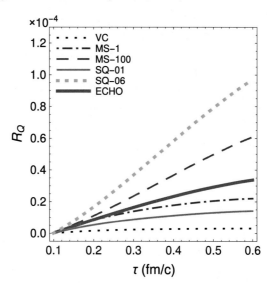

crucial to treat magnetic field as dynamically evolving by properly accounting for medium feedback.

Fig. 6.5 The charge dipole moment ϵ_1^Q for quantifying charge separation as a function of time, computed and compared for various choices of time-dependent magnetic field (with FD initial distribution, $\tau_R = 0.1\,\mathrm{fm/c}$ and $\lambda_5 = 0.2$)

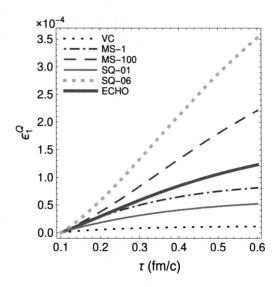

Another way to quantify this CME-induced pre-thermal charge separation is to define a weighed charge dipole moment of the net charge density distribution on the transverse plane, as follows:

$$\epsilon_1^Q = \frac{\int dz dr_\perp^2 d\phi \, r_\perp^2 \, \sin\phi \, n^Q}{\int dz dr_\perp^2 d\phi \, r_\perp^2 \, n^{tot.}} \tag{6.21}$$

where $r_\perp$ and ϕ are transverse radial and azimuthal coordinates on x–y plane, n^Q is the net charge density while $n^{tot.}$ is the total quark and antiquark number density providing the normalization in defining the dimensionless dipole moment ϵ_1^Q. In Fig. 6.5 we show ϵ_1^Q as a function of time. In consistency with R_Q shown in Fig. 6.4, the dipole moment is found to grow with time and sensitively depends on the time evolution of the magnetic field.

So far we have not discussed the influence of two other important ingredients: the initial helicity imbalance parameterized by λ_5 and the peak magnetic field strength B_0. In fact the dependence is fairly simple: the charge separation (described by either R_Q or ϵ_Q^1) is found to be simply linearly proportional to both of these factors. This conclusion should though be put in a context, as the homologous initial chirality imbalance λ_5 used in this work is only an approximate implementation of axial charge dynamics. A dynamical treatment of axial charge would involve inhomogeneous generation from gluon topological fluctuations as well as possible chiral plasma instability [38]. It is worth mentioning that we also studied the influence of the initial momentum distribution as well as the relaxation parameter τ_R on the resulting charge separation dipole, and find that corresponding results are within the same order of magnitude. Detailed results can be found in [39].

It may be noted that the obtained pre-thermal CME-induced charge separation effect has been found to be rather small for our current choice of parameters. This may be due to a number of factors. Firstly, the total evolution time (from 0.1 to 0.6 fm/c) is fairly short, which limits the accumulation of charge separation. Secondly, in the present formulation of chiral kinetic theory the anomalous term responds instantaneously to the applied magnetic field, so the resulting anomalous current decreases in time rapidly along with the magnetic field, thus hindering the buildup of charge separation. This is an important and challenging issue that requires further investigation. Finally, the imbalance in number density between opposite helicity fermions (that we currently use in these calculations) is small compared with the relevant system scale Q_s which is analogous to the situation of a very small ratio for axial charge density n_5 to entropy density in the thermal case.

We end this chapter by discussing the implication for the subsequent hydro-dynamic evolution stage. The occurrence of the pre-hydro CME implies that by the time of hydro onset, the fermion density and current can no longer be set as trivially vanishing (as usually done). Instead, as shown in Figs. 6.2 and 6.3, the per-thermal transport induces a nontrivial initial condition by that time and should be incorporated into the subsequent hydrodynamic evolution as nontrivial initial conditions for the various fermion currents including both the zeroth component (i.e. density) and the spatial component (vector 3-current density). With the AVFD framework, we have tested and demonstrated that indeed such pre-hydro charge separation can be built into hydro initial conditions and will survive through the hydro stage to contribute to the final hadron observables.

With this proof-of-concept study, we aim to develop in the future a more comprehensive and realistic pre-thermal CME modeling tool that will also be seamlessly integrated with an anomalous hydrodynamic evolution. One lesson we've learned from the present study is that the magnetic field driven effects may be significant during the early stage when the field is most strong. For example, the pre-thermal CME leads to nontrivially modified initial conditions for hydrodynamic evolution. In addition to such transport effects, it could be anticipated that the early time magnetic field may have an even stronger and more direct influence on the particles that are dominantly produced through initial hard processes [40, 41], such as the high momentum quarkonia as well as dileptons and photons. During the formation of these particles at the earliest moments after a collision, the strong magnetic field (and the anomalous transport of partons driven by it) may possibly leave an imprint in their production, which would be an interesting topic for future study.

References

1. D.E. Kharzeev, M.A. Stephanov, H.U. Yee, Phys. Rev. **D95**(5), 051901 (2017). https://doi.org/10.1103/PhysRevD.95.051901
2. D.T. Son, N. Yamamoto, Phys. Rev. Lett. **109**, 181602 (2012). https://doi.org/10.1103/PhysRevLett.109.181602
3. D.T. Son, N. Yamamoto, Phys. Rev. **D87**(8), 085016 (2013). https://doi.org/10.1103/PhysRevD.87.085016

4. M.A. Stephanov, Y. Yin, Phys. Rev. Lett. **109**, 162001 (2012). https://doi.org/10.1103/PhysRevLett.109.162001
5. J.Y. Chen, D.T. Son, M.A. Stephanov, H.U. Yee, Y. Yin, Phys. Rev. Lett. **113**(18), 182302 (2014). https://doi.org/10.1103/PhysRevLett.113.182302
6. J.Y. Chen, D.T. Son, M.A. Stephanov, Phys. Rev. Lett. **115**(2), 021601 (2015). https://doi.org/10.1103/PhysRevLett.115.021601
7. J.W. Chen, S. Pu, Q. Wang, X.N. Wang, Phys. Rev. Lett. **110**(26), 262301 (2013). https://doi.org/10.1103/PhysRevLett.110.262301
8. J.H. Gao, Z.T. Liang, S. Pu, Q. Wang, X.N. Wang, Phys. Rev. Lett. **109**, 232301 (2012). https://doi.org/10.1103/PhysRevLett.109.232301
9. J.h. Gao, S. Pu, Q. Wang, Phys. Rev. **D96**(1), 016002 (2017). https://doi.org/10.1103/PhysRevD.96.016002
10. Y. Hidaka, S. Pu, D.L. Yang, Phys. Rev. **D95**(9), 091901 (2017). https://doi.org/10.1103/PhysRevD.95.091901
11. N. Mueller, R. Venugopalan, Phys. Rev. **D97**(5), 051901 (2018). https://doi.org/10.1103/PhysRevD.97.051901
12. E.V. Gorbar, V.A. Miransky, I.A. Shovkovy, P.O. Sukhachov, Phys. Rev. **B95**(20), 205141 (2017). https://doi.org/10.1103/PhysRevB.95.205141
13. M. Mace, S. Schlichting, R. Venugopalan, Phys. Rev. **D93**(7), 074036 (2016). https://doi.org/10.1103/PhysRevD.93.074036
14. M. Mace, N. Mueller, S. Schlichting, S. Sharma, Phys. Rev. **D95**(3), 036023 (2017). https://doi.org/10.1103/PhysRevD.95.036023
15. N. Müller, S. Schlichting, S. Sharma, Phys. Rev. Lett. **117**(14), 142301 (2016). https://doi.org/10.1103/PhysRevLett.117.142301
16. K. Fukushima, Phys. Rev. **D92**(5), 054009 (2015). https://doi.org/10.1103/PhysRevD.92.054009
17. Y. Sun, C.M. Ko, F. Li, Phys. Rev. **C94**(4), 045204 (2016). https://doi.org/10.1103/PhysRevC.94.045204
18. S. Ebihara, K. Fukushima, S. Pu, Phys. Rev. **D96**(1), 016016 (2017). https://doi.org/10.1103/PhysRevD.96.016016
19. D. Vasak, M. Gyulassy, H.T. Elze, Ann. Phys. **173**, 462 (1987). https://doi.org/10.1016/0003-4916(87)90169-2
20. P. Zhuang, U.W. Heinz, Ann. Phys. **245**, 311 (1996). https://doi.org/10.1006/aphy.1996.0011
21. P.f. Zhuang, U.W. Heinz, Phys. Rev. **D53**, 2096 (1996). https://doi.org/10.1103/PhysRevD.53.2096
22. P.f. Zhuang, U.W. Heinz, Phys. Rev. **D57**, 6525 (1998). https://doi.org/10.1103/PhysRevD.57.6525
23. S. Ochs, U.W. Heinz, Ann. Phys. **266**, 351 (1998). https://doi.org/10.1006/aphy.1998.5796
24. X. Guo, P. Zhuang, Phys. Rev. **D98**(1), 016007 (2018). https://doi.org/10.1103/PhysRevD.98.016007
25. S.R. De Groot, *Relativistic Kinetic Theory. Principles and Applications* (North-holland, Amsterdam, 1980), 417p
26. A. Huang, S. Shi, Y. Jiang, J. Liao, P. Zhuang, Phys. Rev. **D98**(3), 036010 (2018). https://doi.org/10.1103/PhysRevD.98.036010
27. H. Kowalski, T. Lappi, R. Venugopalan, Phys. Rev. Lett. **100**, 022303 (2008). https://doi.org/10.1103/PhysRevLett.100.022303
28. J.P. Blaizot, F. Gelis, J.F. Liao, L. McLerran, R. Venugopalan, Nucl. Phys. **A873**, 68 (2012). https://doi.org/10.1016/j.nuclphysa.2011.10.005
29. J.P. Blaizot, J. Liao, L. McLerran, Nucl. Phys. **A920**, 58 (2013). https://doi.org/10.1016/j.nuclphysa.2013.10.010
30. J.P. Blaizot, B. Wu, L. Yan, Nucl. Phys. **A930**, 139 (2014). https://doi.org/10.1016/j.nuclphysa.2014.07.041
31. W.T. Deng, X.G. Huang, Phys. Rev. **C85**, 044907 (2012). https://doi.org/10.1103/PhysRevC.85.044907

32. L. McLerran, V. Skokov, Nucl. Phys. **A929**, 184 (2014). https://doi.org/10.1016/j.nuclphysa. 2014.05.008
33. Y. Jiang, S. Shi, Y. Yin, J. Liao, Chin. Phys. **C42**(1), 011001 (2018). https://doi.org/10.1088/ 1674-1137/42/1/011001
34. Y. Yin, J. Liao, Phys. Lett. **B756**, 42 (2016). https://doi.org/10.1016/j.physletb.2016.02.065
35. H.U. Yee, Y. Yin, Phys. Rev. **C89**(4), 044909 (2014). https://doi.org/10.1103/PhysRevC.89. 044909
36. G. Inghirami, L. Del Zanna, A. Beraudo, M.H. Moghaddam, F. Becattini, M. Bleicher, Eur. Phys. J. **C76**(12), 659 (2016). https://doi.org/10.1140/epjc/s10052-016-4516-8
37. J. Bloczynski, X.G. Huang, X. Zhang, J. Liao, Phys. Lett. **B718**, 1529 (2013). https://doi.org/ 10.1016/j.physletb.2012.12.030
38. Y. Akamatsu, N. Yamamoto, Phys. Rev. Lett. **111**, 052002 (2013). https://doi.org/10.1103/ PhysRevLett.111.052002
39. A. Huang, Y. Jiang, S. Shi, J. Liao, P. Zhuang, Phys. Lett. **B777**, 177 (2018). https://doi.org/ 10.1016/j.physletb.2017.12.025
40. X. Guo, S. Shi, N. Xu, Z. Xu, P. Zhuang, Phys. Lett. **B751**, 215 (2015). https://doi.org/10. 1016/j.physletb.2015.10.038
41. G. Basar, D.E. Kharzeev, E.V. Shuryak, Phys. Rev. **C90**(1), 014905 (2014). https://doi.org/ 10.1103/PhysRevC.90.014905

Chapter 7
Rotation of the QCD Plasma and the Chiral Vortical Effect

In previous chapters we perform detail and quantitative study on the Chiral Magnetic Effect in relativistic heavy-ion collisions. The CME provides a special opportunity to observe the effect of $\mathcal{P}$- and $\mathcal{CP}$-violating axial charge, an interplay of topological charge transition and chiral anomaly. On the other hand, in heavy-ion collisions there exists another global axial vector field, the vorticity field, which couples to the (pseudo-scalar) axial charge, and generates a vector current that could be observed. As the analog to the Chiral Magnetic Effect and Chiral Magnetic Wave, these anomalous chiral transport effects induced by strong fluid rotation are named as Chiral Vortical Effect and Chiral Vortical Wave [1–3].

It is worth noting that while the time evolution of the magnetic field remains to be one of the biggest theoretical uncertainties in quantitative studies of Chiral Magnetic Effects, the space-time profile of the vorticity field could be studied in great depth, by using either hydrodynamic simulations or other transport models, with realistic initial conditions. In the meanwhile, due to conservation of angular momentum, the vorticity field is expected to have longer lifetime and be able to provide more stable driving force of chiral transportation.

In this chapter, we first focus on the vorticity profile of the QCD Plasma created in heavy-ion collisions, and corresponding direct measurements of such global rotation; then we discuss how one can measure vortical effects induced by strong fluid rotation.

7.1 Vorticity in Heavy-Ion Collisions

The study of strongly interacting matter under rotation has attracted significant interest recently across disciplines such as condensed matter physics, cold atomic gases, astrophysics, and nuclear physics [4–12]. Besides aforementioned CVE and CVW, the rotation is also found to influence the phase structures and transitions

© Springer Nature Switzerland AG 2019
S. Shi, *Soft and Hard Probes of QCD Topological Structures
in Relativistic Heavy-Ion Collisions*, Springer Theses,
https://doi.org/10.1007/978-3-030-25482-7_7

in various physical systems [13–16]. In heavy-ion collision experiments, currently carried out at the Relativistic Heavy Ion Collider (RHIC) and the Large Hadron Collider (LHC), it has long been expected that the large angular momentum carried out by the colliding system might lead to observable effects such as global polarization of certain produced particles [17–22].

Here we focus on the case of heavy-ion collisions, in which an extremely hot subatomic material known as a quark–gluon plasma (QGP) is created. The QGP was the form of matter in the early Universe moments after the Big Bang, and is now recreated in laboratory by RHIC and the LHC. The created hot material has been found to undergo a strong collective expansion that is well described by relativistic hydrodynamics. In a typical non-central collision, the two opposite-moving nuclei have their center-of-mass misaligned and thus carry a considerable angular momentum J_y of about $10^{4\sim5}\hbar$ along the out-of-plane direction, with a quick estimate $J_y \sim bA\sqrt{s}/2$ where $b, A, \sqrt{s}$ are impact parameter, nuclear mass number, and colliding beam energy. It has been long believed that a good fraction of this angular momentum will remain in the crated hot QGP and lead to strong nonzero vortical fluid structures (often quantified by fluid vorticity) during its hydrodynamic evolution, thus forming the "subatomic swirls." Indeed, many quantitative computational results have suggested their existence [23–30], awaiting an experimental verification.

Recently the STAR Collaboration reported their remarkable measurement of the "subatomic swirls" in the AuAu collisions at RHIC [31]. A possible signature of the fluid vorticity is the spin polarization of the produced particles which on average should be aligned with the colliding system's global angular momentum direction. But this is an extremely challenging type of measurement. By a clever analysis of spin orientation of the produced subatomic particles called Λ hyperons, the STAR Collaboration was able to find very strong evidence for the global polarization effect, from which they extracted an average fluid vorticity of about $10^{21}\,\mathrm{s}^{-1}$, being the most vortical fluid ever known.

For such an important finding, it is crucial to look for additional evidences of confirmation and to critically test current interpretations of the polarization data. In this study we propose to use the CuCu and CuAu colliding systems at RHIC as an ideal and natural way to provide the necessary verification. Such experiments were performed previously at RHIC, and a number of measurements were previously carried out at several beam energies by PHENIX, PHOBOS, and STAR [32–35]. We will compute the vorticity structures in the CuCu and CuAu systems, to be compared with the AuAu system. Interestingly, our calculations will show that the fluid vorticity in the CuCu or CuAu collision has a similar pattern and is comparable in magnitude to that in the AuAu Collision. We make quantitative predictions for the polarization measurements versus the collisional beam energy, which can be readily tested by experimental data.

7.1.1 The Transport Model Setup

The rotation of a system is quantified by vorticity. One could define the vorticity tensor as the curl of the fluid velocity u^μ:

$$\Omega_{\mu\nu} = \frac{1}{2}(\partial_\nu u_\mu - \partial_\mu u_\nu) , \qquad (7.1)$$

while the vorticity vector field as

$$\Omega^\mu = \frac{1}{2}\epsilon^{\mu\nu\rho\sigma} u_\nu \Omega_{\rho\sigma} . \qquad (7.2)$$

In the non-relativistic limit, it becomes the familiar three-vorticity $\omega = \frac{1}{2}\nabla \times \mathbf{v}$ with $\mathbf{v}$ the three-velocity. To see the physical meaning of vorticity, let us consider the non-relativistic fluid moving as rigid body, i.e. $\mathbf{v} = \mathbf{\Omega} \times \mathbf{r}$, and one can easily find its vorticity $\omega = \mathbf{\Omega}$ at any point! The vorticity reflects the angular velocity of the local rotation, while a global vorticity field reflects the global rotation of the system.

In this study, we use a widely adopted transport model for heavy-ion collisions, the AMPT (A Multi-Phase Transport) model [36–40]. In particular we use the string melting version of the AMPT model [36, 37] which includes the initial particle production right after the primary collision of the two incoming nuclei, an elastic parton cascade, a quark coalescence model for hadronization, and a hadronic cascade. The AMPT model provides a very reasonable description of the bulk evolution. We will use the same set of parameters as in [38], where the simulations with those parameters well reproduced the yields, transverse momentum spectra and v_2 data for low-p_T pions and kaons in central and mid-central Au+Au collisions at RHIC. To be specific, the parameters include the Lund string fragmentation parameters ($a = 0.55$, $b = 0.15/\text{GeV}^2$), strong coupling constant $\alpha_s = 0.33$ for the parton cascade, a parton cross-section of 3 mb (i.e. a parton Debye screening mass $\mu = 2.265/\text{fm}$), and an upper limit of 0.40 on the relative production of strange to nonstrange quarks.

An advantage of the AMPT model is that it allows explicit tracking of every parton or hadron's motion during the evolution. This allows a relatively straightforward extraction of the system's angular momentum as well as fluid rotation. The model was first used in [27] to compute the structures of local fluid vorticity as in Eq. (7.1), where the four-velocity of fluid cells u^μ is computed in AMPT from averaging thousands of events for a given collision energy and impact parameter. The interesting component is the one along the out-of-plane direction $\Omega_{3,1}$, which in the non-relativistic limit becomes the familiar y-component ω_y of the three-vorticity $\omega = \frac{1}{2}\nabla \times \mathbf{v}$ with $\mathbf{v}$ the three-velocity. The global rotation effect can be quantified by properly averaging over the whole fireball. The results for AuAu collisions obtained in [27] predicted a strong monotonic decrease of the average vorticity with increasing beam energy, in consistency with later experimental data [31]. Another advantage of the AMPT model is that the finally observed hadrons are explicitly formed via quark coalescence. It is relatively easy to incorporate the spin polarization effect upon the formation of hadrons such as the Λ hyperon. More specifically, the ensemble-averaged spin 4-vector of the produced Λ is determined from the local vorticity at its formation location, as [30, 31, 41]:

$$S^\mu \equiv -\frac{1}{8mT}\epsilon^{\mu\nu\rho\sigma}p_\nu\Omega_{\rho\sigma} \tag{7.3}$$

where p^ν is the four-momentum of the hyperon and m the hyperon mass. Such polarization could then be properly converted to the P_Λ observable measured by the STAR in [31]. A subsequent AMPT study [41] for AuAu collisions included the polarization due to local vorticity for each produced hadron in the AMPT framework to calculate the overall spin polarization projected onto the global angular momentum direction. The results for Λ polarization are in agreement with the STAR data. It should however be emphasized that, strictly speaking, in a relativistic fluid it is the thermal vorticity that determines the particle polarization: see a detailed analysis in [22]. The thermal vorticity differs from the vorticity in Eq. (7.1) by terms containing gradients of temperature $\sim\partial_\mu(T)$, which is straightforward to evaluate in hydrodynamic calculations but less obvious in transport models. The influence of this theoretical uncertainty requires future investigation and could be calibrated by comparison with experimental data in AuAu collisions (see below).

A versatile aspect of the present model is that it allows easy adaptation to be used for studying various different colliding systems across a wide beam energy span. In this study, we use the same AMPT model setup and computing methods as that in [27], albeit for the investigations of two different systems: the CuCu and CuAu collisions. Our goal is to compute the fluid vorticity in these systems and to make quantitative predictions for the polarization measurements versus the collisional beam energy, which can be readily tested by future experimental analysis. As a step of model validation, we first use the AMPT to compute the Λ polarization in AuAu collisions. Note that in the calculation we use the same centrality class and kinematic selection as the STAR measurement. Furthermore based on the estimates by STAR [31], we've taken into account the 20% suppression effect on top of the primary Λ polarization directly computed from the model due to the hyperons from resonance decays. The results, shown in Fig. 7.1, are consistent with results from [41] and compare well with the STAR data. This demonstrates that predictions from this approach are quantitatively robust. It may be noted that experimental data indicate at a difference between P_Λ and $P_{\bar\Lambda}$ which might be due to their opposite magnetic moments provided there is a magnetic field [30]. The current transport model does not include any magnetic field and only includes the rotational polarization effect which does not distinguish particle or antiparticle and therefore predicts the same polarization for Λ and $\bar\Lambda$.

Fig. 7.1 The Λ hyperon global polarization P_H results for AuAu collisions at RHIC, computed from the AMPT model (see text) and compared with the STAR data

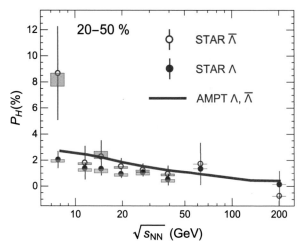

7.1.2 Fluid Vorticity and Particle Polarization for Cu–Cu and Cu–Au Collisions

Let us now present the new results for CuCu and CuAu collisions, in comparison with AuAu collisions. As already demonstrated in previous works [27–29], the main contributions to the $\Omega_{3,1}$ arise from the nontrivial distribution patterns of the fluid velocity vectors on the Reaction Plane, i.e. the x–z or x–η plane. Note we use the (τ, η, x, y) coordinates to present results with τ and η the proper time and spatial rapidity as widely used in heavy ion phenomenology. To give an intuitive picture of the vorticity structures in these different systems, we first show in Fig. 7.2 the velocity component distribution $\mathbf{v} = (v_x, v_\eta)$ on the x–η plane as well as the corresponding vorticity component $\Omega_{3,1}$ distribution. The three panels are for AuAu, CuAu, and CuCu collisions, respectively, all computed for 20–50% centrality at $\sqrt{s} = 63$ GeV at time $t = 2$ fm/c. While we've done calculations at a variety of beam energies, we choose this particular energy for comparison as experimental data were taken at this energy for all three systems. From Fig. 7.2, one sees that the vorticity distribution patterns, consistent with previous AuAu results [27], are highly similar among these different systems. We find that despite the difference in system size, spatial spread of fireball, and total angular momentum, even the absolute magnitude of vorticity component $\Omega_{3,1}$ is very close for all of them. We've verified this observation to be true for calculations at varied values of the collisional beam energy.

In order to more quantitatively compare the systems, we compute the averaged vorticity component $\Omega_{3,1}$ as a function of time. While the local vorticity $\Omega_{3,1}$ varies a lot and even oscillates in sign across the fireball (as seen in Fig. 7.2), after properly averaging over the fireball, the obtained $\langle \Omega_{3,1} \rangle$ has a definitive sign that aligns with the global angular momentum and is directly relevant for the size of the observable polarization effect in the end. We use the same averaging procedure as that used

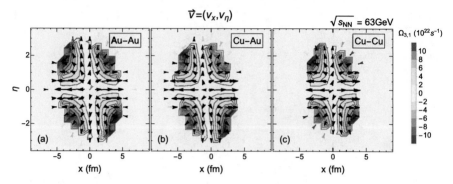

Fig. 7.2 The velocity component distribution $\mathbf{v} = (v_x, v_\eta)$ (black arrows) on the x–η plane and the corresponding vorticity component $\Omega_{3,1}$ distribution (colored contour plot), for 20–50% AuAu, CuAu and CuCu collisions at $\sqrt{s_{NN}} = 63$ GeV at time $t = 2$ fm/c

Fig. 7.3 The averaged vorticity $\langle\Omega_{3,1}\rangle$ as a function of time, for 20–50% AuAu, CuAu and CuCu collisions at $\sqrt{s_{NN}} = 63$ GeV respectively

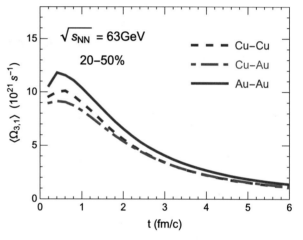

for the AuAu calculations in [27] to compute the $\langle\Omega_{3,1}\rangle$ for the CuAu and CuCu systems. The results are shown in Fig. 7.3. Indeed the values of $\langle\Omega_{3,1}\rangle$ are found to be fairly comparable for all of them with slightly larger values for the AuAu system, confirming the qualitative observation from Fig. 7.2. This quantitative comparison also implies that the potentially observable signals, i.e. the particle polarization effects, should also be comparable, thus equally measurable for CuAu and CuCu systems as well. It should be noted that due to their difference in system size, each colliding system has a somewhat different intrinsic evolution time scale. Therefore the comparison in Fig. 7.3 at the same time should be put in such context.

To verify such expectation, we then compute the Λ hyperon polarization, in the same way as the calculation for AuAu (which is validated with data in Fig. 7.1). The results for the P_Λ versus collisional beam energy $\sqrt{s_{NN}}$ are shown in Fig. 7.4 for 20–50% AuAu, CuAu, and CuCu collisions, respectively. While the predicted signals are indeed comparable, we find a very interesting hierarchy of the signal

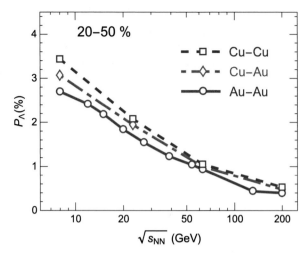

Fig. 7.4 The predicted P_Λ signals versus collisional beam energy $\sqrt{s_{NN}}$ for 20–50% CuCu and CuAu collisions in comparison with the AuAu collisions

strength: CuCu > CuAu > AuAu. At first sight, this appears counterintuitive, especially given that the averaged vorticity is slightly bigger in AuAu system as shown in Fig. 7.3. Upon careful examination, it is found that the key to understand this hierarchy is the timing of the hyperon production. Because CuCu system is much diluter than the CuAu system which is still diluter than the AuAu system, the fireball evolution time till freeze-out is generally shorter. As a result, the Λ hyperons are on average formed earlier in CuCu system while later in AuAu system. This statement has been explicitly checked in our numerical calculations: the Λ production rate curve versus evolution time is visibly shifted to the earlier time side than that in the AuAu collisions. This timing effect, put in the perspective of the rapid decrease of $\langle \Omega_{3,1} \rangle$ with time in a similar fashion for all three systems (shown in Fig. 7.3), implies that the Λ produced in CuCu system is on average influenced by a larger vorticity value, thus shows stronger polarization effect, as compared with that in AuAu system. Our predictions for the hyperon polarization in CuCu and CuAu systems and for the interesting hierarchy in the signal strength could be readily tested with experimental analyses.

In summary, we've used a transport model to investigate the fluid vorticity structures as well as the global hyperon polarization effect in two new colliding systems: the CuCu and CuAu collisions. A detailed picture of the vorticity structures in those systems was obtained and found to be very similar to that in AuAu collisions at the same beam energy and centrality class. The average vorticity component along the out-of-plane direction was quantitatively extracted and also found to be comparable with and has a similar time dependence as that in AuAu collisions. A most unexpected finding comes from the computed final state hadron observable, namely the Λ hyperon polarization, where a stronger signal is found in the CuCu and CuAu systems compared to the AuAu system. The interesting hierarchy of polarization signal (CuCu > CuAu > AuAu) could be understood from the time dependence of vorticity evolution and the different timing for the hyperon production in these systems. These predictions can be readily tested by experimental data, and we propose the global hyperon polarization measurements in CuCu and CuAu collisions as an ideal and independent verification for the subatomic swirl discovery and for our current interpretation of the observed signal as rotational polarization effect.

7.2 Chiral Vortical Effects

Once the fluid vorticity structures are well established, one would be really to explore the vorticity-driven anomalous chiral transport effects in all these systems. Similarly to the efforts for studying Chiral Magnetic Effects, the quantitative study of Chiral Vortical Effects would require the future development of hydrodynamic frameworks that incorporate both the 3D fluid vorticity structures and the anomalous transport currents required by chiral anomaly.

To quantify the Chiral Vortical Effects, one can start from the chiral vortical current, which is an analog to the CME currents,

$$\hat{D}_\mu J^\mu_{\chi,f} = 0 \tag{7.4}$$

$$J^\mu_{\chi,f} = n_{\chi,f} u^\mu + v^\mu_{\chi,f} + \chi N_c \left(\frac{T^2}{12} + \frac{(\mu_{\chi,f})^2}{4\pi^2} \right) \Omega^\mu \tag{7.5}$$

$$\Delta^\mu_\nu \hat{d} \left(v^\nu_{\chi,f} \right) = -\frac{1}{\tau_r} \left[\left(v^\mu_{\chi,f} \right) - \left(v^\mu_{\chi,f} \right)_{NS} \right] \tag{7.6}$$

$$\left(v^\mu_{\chi,f} \right)_{NS} = \frac{\sigma}{2} T \Delta^{\mu\nu} \partial_\nu \left(\frac{\mu_{\chi,f}}{T} \right). \tag{7.7}$$

unlike CME which generates extra electric current, the CVE depends on baryonic number rather than electric charges, hence it generates nonzero baryonic current

$$J_B = \frac{N_f N_c}{3\pi^2} \mu_B \mu_{B,5} \omega, \tag{7.8}$$

and causes baryon separations. By measuring the two-baryon azimuthal correlation,

$$\gamma^{\alpha\beta}_B \equiv \langle \cos(\phi^\alpha + \phi^\beta - 2\psi_{EP}) \rangle \tag{7.9}$$

$$\delta^{\alpha\beta}_B \equiv \langle \cos(\phi^\alpha - \phi^\beta) \rangle, \tag{7.10}$$

where α, β denote for baryons/anti-baryons. With precise experimental measurements on two-baryon azimuthal correlations in future, the CVE observables are expected to be independent probes of the vorticity profile and the topological charge transition in relativistic heavy-ion collisions.

References

1. D.T. Son, P. Surowka, Phys. Rev. Lett. **103**, 191601 (2009). https://doi.org/10.1103/PhysRevLett.103.191601
2. D.E. Kharzeev, D.T. Son, Phys. Rev. Lett. **106**, 062301 (2011). https://doi.org/10.1103/PhysRevLett.106.062301
3. Y. Jiang, X.G. Huang, J. Liao, Phys. Rev. **D92**(7), 071501 (2015). https://doi.org/10.1103/PhysRevD.92.071501
4. J. Gooth et al., Nature **547**, 324 (2017). https://doi.org/10.1038/nature23005
5. A.L. Fetter, Rev. Mod. Phys. **81**, 647 (2009). https://doi.org/10.1103/RevModPhys.81.647
6. M. Urban, P. Schuck, Phys. Rev. A **78**(1), 011601 (2008). https://doi.org/10.1103/PhysRevA.78.011601
7. M. Iskin, E. Tiesinga, Phys. Rev. A **79**(5), 053621 (2009). https://doi.org/10.1103/PhysRevA.79.053621
8. E. Berti, F. White, A. Maniopoulou, M. Bruni, Mon. Not. R. Astron. Soc. **358**, 923 (2005). https://doi.org/10.1111/j.1365-2966.2005.08812.x
9. A.L. Watts et al., Rev. Mod. Phys. **88**(2), 021001 (2016). https://doi.org/10.1103/RevModPhys.88.021001
10. I.A. Grenier, A.K. Harding, C. R. Phys. **16**, 641 (2015). https://doi.org/10.1016/j.crhy.2015.08.013
11. D.E. Kharzeev, J. Liao, S.A. Voloshin, G. Wang, Prog. Part. Nucl. Phys. **88**, 1 (2016). https://doi.org/10.1016/j.ppnp.2016.01.001
12. J. Liao, Pramana **84**(5), 901 (2015). https://doi.org/10.1007/s12043-015-0984-x
13. Y. Jiang, J. Liao, Phys. Rev. Lett. **117**(19), 192302 (2016). https://doi.org/10.1103/PhysRevLett.117.192302
14. S. Ebihara, K. Fukushima, K. Mameda, Phys. Lett. **B764**, 94 (2017). https://doi.org/10.1016/j.physletb.2016.11.010
15. H.L. Chen, K. Fukushima, X.G. Huang, K. Mameda, Phys. Rev. **D93**(10), 104052 (2016). https://doi.org/10.1103/PhysRevD.93.104052
16. M.N. Chernodub, S. Gongyo, Phys. Rev. **D95**(9), 096006 (2017). https://doi.org/10.1103/PhysRevD.95.096006
17. Z.T. Liang, X.N. Wang, Phys. Rev. Lett. **94**, 102301 (2005). https://doi.org/10.1103/PhysRevLett.94.102301, https://doi.org/10.1103/PhysRevLett.96.039901. [Erratum: Phys. Rev. Lett.96,039901(2006)]
18. J.H. Gao, S.W. Chen, W.t. Deng, Z.T. Liang, Q. Wang, X.N. Wang, Phys. Rev. **C77**, 044902 (2008). https://doi.org/10.1103/PhysRevC.77.044902
19. S.A. Voloshin (2004). arXiv:nucl-th/0410089
20. B. Betz, M. Gyulassy, G. Torrieri, Phys. Rev. **C76**, 044901 (2007). https://doi.org/10.1103/PhysRevC.76.044901
21. F. Becattini, F. Piccinini, J. Rizzo, Phys. Rev. **C77**, 024906 (2008). https://doi.org/10.1103/PhysRevC.77.024906
22. F. Becattini, V. Chandra, L. Del Zanna, E. Grossi, Annals Phys. **338**, 32 (2013). https://doi.org/10.1016/j.aop.2013.07.004
23. F. Becattini, L. Csernai, D.J. Wang, Phys. Rev. **C88**(3), 034905 (2013). https://doi.org/10.1103/PhysRevC.93.069901, https://doi.org/10.1103/PhysRevC.88.034905. [Erratum: Phys. Rev.C93,no.6,069901(2016)]
24. L.P. Csernai, V.K. Magas, D.J. Wang, Phys. Rev. **C87**(3), 034906 (2013). https://doi.org/10.1103/PhysRevC.87.034906
25. L.P. Csernai, D.J. Wang, M. Bleicher, H. Stöcker, Phys. Rev. **C90**(2), 021904 (2014). https://doi.org/10.1103/PhysRevC.90.021904
26. F. Becattini, G. Inghirami, V. Rolando, A. Beraudo, L. Del Zanna, A. De Pace, M. Nardi, G. Pagliara, V. Chandra, Eur. Phys. J. **C75**(9), 406 (2015). https://doi.org/10.1140/epjc/s10052-015-3624-1

27. Y. Jiang, Z.W. Lin, J. Liao, Phys. Rev. **C94**(4), 044910 (2016). https://doi.org/10.1103/PhysRevC.94.044910, https://doi.org/10.1103/PhysRevC.95.049904. [Erratum: Phys. Rev.C95,no.4,049904(2017)]
28. W.T. Deng, X.G. Huang, Phys. Rev. **C93**(6), 064907 (2016). https://doi.org/10.1103/PhysRevC.93.064907
29. L.G. Pang, H. Petersen, Q. Wang, X.N. Wang, Phys. Rev. Lett. **117**(19), 192301 (2016). https://doi.org/10.1103/PhysRevLett.117.192301
30. F. Becattini, I. Karpenko, M. Lisa, I. Upsal, S. Voloshin, Phys. Rev. **C95**(5), 054902 (2017). https://doi.org/10.1103/PhysRevC.95.054902
31. L. Adamczyk et al., Nature **548**, 62 (2017). https://doi.org/10.1038/nature23004
32. A. Adare et al., Phys. Rev. **C78**, 044902 (2008). https://doi.org/10.1103/PhysRevC.78.044902
33. B. Alver et al., Phys. Rev. **C83**, 024913 (2011). https://doi.org/10.1103/PhysRevC.83.024913
34. A. Adare et al., Phys. Rev. **C94**(5), 054910 (2016). https://doi.org/10.1103/PhysRevC.94.054910
35. L. Adamczyk et al., Phys. Rev. Lett. **118**(1), 012301 (2017). https://doi.org/10.1103/PhysRevLett.118.012301
36. Z.W. Lin, C.M. Ko, B.A. Li, B. Zhang, S. Pal, Phys. Rev. **C72**, 064901 (2005). https://doi.org/10.1103/PhysRevC.72.064901
37. Z.w. Lin, C.M. Ko, Phys. Rev. **C65**, 034904 (2002). https://doi.org/10.1103/PhysRevC.65.034904
38. Z.W. Lin, Phys. Rev. **C90**(1), 014904 (2014). https://doi.org/10.1103/PhysRevC.90.014904
39. G.L. Ma, B. Zhang, Phys. Lett. **B700**, 39 (2011). https://doi.org/10.1016/j.physletb.2011.04.057
40. Q.Y. Shou, G.L. Ma, Y.G. Ma, Phys. Rev. **C90**(4), 047901 (2014). https://doi.org/10.1103/PhysRevC.90.047901
41. H. Li, L.G. Pang, Q. Wang, X.L. Xia, Phys. Rev. **C96**(5), 054908 (2017). https://doi.org/10.1103/PhysRevC.96.054908

Part II
Hard Probe of the
Chromo-Magnetic-Monopoles

Chapter 8
Jet Energy-Loss Simulations

As mentioned in Sect. 1.3.3, hard particles with $p_T \gtrsim 10$ GeV are produced from initial hard processes. They interact with the hot medium with coupling in perturbative regime. High energy jets provide invaluable tomographic "imaging" of the hot bulk medium created in such collision. By comparing jet observables from AA collisions with that of pp collisions, one can learn about the properties of the hot medium created in high energy nuclear collisions, as well as how jets interact with the medium. A conventional observable for quantifying medium attenuation of jets is the high p_T hadron's nuclear modification factor R_{AA} defined as:

$$R_{AA}(p_T; \phi) = R_{AA}(p_T) [1 + 2v_2(p_T) \cos(2\phi - 2\Psi_2)$$
$$+ 2v_3(p_T) \cos(3\phi - 3\Psi_3) + \ldots] , \qquad (8.1)$$

where the coefficient v_n in the angular distribution of the jet fragments is essentially measured with respect to azimuthal flow harmonics $v_n^{\text{soft}} e^{i\Psi_n}$ of the soft ($p_T < 2$ GeV) hadronic fragments from the QCD fluid.

Strong suppression effect (with R_{AA} significantly less than one) as well as sizable anisotropy v_2 at high transverse momentum have been consistently observed from RHIC to LHC. There are now comprehensive sets of available data, from average suppression to azimuthal anisotropy, from light to heavy flavors, from RHIC 200 GeV to LHC 2.76 TeV as well as 5.02 TeV collisions. A unified description of such comprehensive data presents a stringent vetting of any viable model for jet quenching phenomenology, as indeed demonstrated by past studies. For example the azimuthal anisotropy was found to pose severe challenge for models and hint at highly nontrivial temperature dependence of jet-medium coupling [1–4]. The beam-energy dependence was also found to suggest a considerable reduction of average medium opaqueness from RHIC to LHC (which again hints at strong temperature dependence) [2, 3, 5–7]. The puzzling close proximity between light flavor and

© Springer Nature Switzerland AG 2019
S. Shi, *Soft and Hard Probes of QCD Topological Structures
in Relativistic Heavy-Ion Collisions*, Springer Theses,
https://doi.org/10.1007/978-3-030-25482-7_8

heavy flavor energy loss was found to indicate at the necessity of including elastic energy loss and magnetic screening effect [8, 9].

To quantitatively understand jet-quenching observables in both RHIC and LHC experiments, one requires adequate modeling for three different components: (1) what is the distribution for initially produced hard partons; (2) how hard partons interact with the soft medium and loss energy; and finally (3) how a final state hard parton fragments into colorless hadrons. It shall be noted that the (1) and (3) component, i.e. p_T-differential cross-section for initial hard partons and fragmentation functions (FF's) shall be identical in AA collisions and pp collisions, and the convolution of these two components shall give data comparable hadron p_T spectra in pp collisions:

$$\frac{d}{dp_T}\sigma^{pp}_{hadron}(p_T) = \sum_{parton} \frac{d}{dp}\sigma^{pp}_{parton}(p) \otimes D(p \to p_T) . \tag{8.2}$$

However, in AA collisions, the final parton p_T-spectra would be modified as the quenching effect,

$$\frac{d}{dp_T}\sigma^{AA}_{hadron}(p_T, \phi) = \sum_{parton} \frac{d}{dp}\sigma^{pp}_{parton}(p) \otimes P(p, \phi \to p_f, \phi)$$
$$\otimes D(p_f, \phi \to p_T, \phi) . \tag{8.3}$$

Given that the effective medium path length depends on the jet direction, we expect nontrivial azimuthal angular dependence of hadron spectra in AA collisions.

In the CUJET frameworks, the pp spectra of light quarks and gluons are generated by LO pQCD (X.N. Wang, private communication) calculations with CTEQ5 Parton Distribution Functions; while those of charm and bottom quarks are generated from FONLL calculation [10] with CTEQ6M Parton Distribution Functions. In the meanwhile, the spectra of light hadrons are computed with KKP FF's [11]; and those of open heavy flavor mesons computed Peterson FF's [12] (taking $\epsilon = 0.06$ for D meson, and $\epsilon = 0.006$ for B meson). The decay of heavy flavor mesons into leptons, including $D \to \ell$, $B \to \ell$, and $B \to D \to \ell$ channels, follows the same parameterization as in [10].

With all these components as reliable input, the CUJET3 [13, 14] model is a jet energy loss simulation framework including both radiational and elastic collisional energy loss, built upon a non-perturbative microscopic model for the hot medium, as a semi-Quark-Gluon-Monopole plasma (sQGMP) which integrates two essential elements of confinement, i.e. the suppression of quarks/gluons and emergent magnetic monopoles. Besides CUJET3, currently there are a number of jet energy loss modeling frameworks differing in their implementation of hot medium and energy loss scheme with varied degrees of sophistication [15–19], and the large amount of high precision data will be a great opportunity to quantitatively analyze the phenomenological viability of each framework.

8.1 CUJET3 Framework

The CUJET(x) simulation packages are a series of developing packages for jet energy loss modeling. Its first version, referred to as CUJET1.0 [8], implements the DGLV radiational process and TG elastic scattering energy loss and couples with a simplified bulk background, and correctly explains the different energy loss mechanism between heavy and light flavors. After that, the next version, CUJET2.0 [9], couples with more realistic bulk background provided by smooth VISH2+1 hydrodynamic simulations, provides more precise description of R_{AA}, and predicts non-vanishing anisotropy v_2. Both CUJET1.0 and CUJET2.0 models assume only chromo-electric components—quarks and gluons—in the medium. Eventually, the CUJET3.0 [13, 14] framework includes the chromo-magnetic-monopole degree of freedom. It is built upon a non-perturbative microscopic model for the hot medium as a semi-Quark-Gluon-Monopole Plasma (sQGMP) which integrates two essential elements of confinement, i.e. the suppression of quarks/gluons and emergent magnetic monopoles. With enhanced energy-loss at the near T_c regime, the CUJET3.0 model provides correct predictions for both R_{AA} and v_2 observables.

Explicitly, the CUJET3 model employs TG elastic energy loss formula [20–22] for collisional processes, with energy loss

$$\frac{dE(z)}{d\tau} = -C_R\pi\left[\alpha_s(\mu(z))\alpha_s(6E(z)\Gamma(z)T(z))\right]T(z)^2\left(1+\frac{N_f}{6}\right)$$
$$\times \log\left[\frac{6T(z)\sqrt{E(z)^2\Gamma(z)^2-M^2}}{\left(E(z)\Gamma(z)-\sqrt{E(z)^2\Gamma(z)^2-M^2}+6T(z)\right)\mu(z)}\right], \tag{8.4}$$

and the average number of collisions

$$\bar{N}_c = \int_0^{\tau_{max}} d\tau \left[\frac{\alpha(\mu(z))\alpha(6E(z)\Gamma(z)T(z))}{\mu(z)^2}\right]\left[\frac{\Gamma(z)}{\gamma_f}\frac{18\zeta(3)}{\pi}(4+N_f)T(z)^3\right], \tag{8.5}$$

where the $E(z)$ integral equation is solved recursively. For radiational processes, the CUJET3 model employs the dynamical DGLV opacity expansion theory [23–26] with Shuryak–Liao chromo-magnetic-monopole scenario [1, 27–31]. The inclusive single gluon emission spectrum in the $n=1$ opacity series reads:

$$x_E\frac{dN_g^{n=1}}{dx_E} = \frac{18C_R}{\pi^2}\frac{4+N_f}{16+9N_f}\int d\tau\,\rho(z)\Gamma(z)\int d^2k_\perp\alpha_s\left(\frac{k_\perp^2}{x_+(1-x_+)}\right)$$
$$\times \int d^2q \frac{\alpha_s^2(q_\perp^2)\left(f_E^2+\frac{f_E^2f_M^2\mu^2(z)}{q_\perp^2}\right)\chi_T+\left(f_M^2+\frac{f_E^2f_M^2\mu^2(z)}{q_\perp^2}\right)(1-\chi_T)}{(q_\perp^2+f_E^2\mu^2(z))(q_\perp^2+f_M^2\mu^2(z))}$$

$$\times \frac{-2(\mathbf{k}_\perp - \mathbf{q}_\perp)}{(\mathbf{k}_\perp - \mathbf{q}_\perp)^2 + \chi^2(z)} \left[\frac{\mathbf{k}_\perp}{\mathbf{k}_\perp^2 + \chi^2(z)} - \frac{(\mathbf{k}_\perp - \mathbf{q}_\perp)}{(\mathbf{k}_\perp - \mathbf{q}_\perp)^2 + \chi^2(z)} \right]$$

$$\times \left[1 - \cos\left(\frac{(\mathbf{k}_\perp - \mathbf{q}_\perp)^2 + \chi^2(z)}{2x_+ E} \tau \right) \right] \left(\frac{x_E}{x_+} \right) \left| \frac{dx_+}{dx_E} \right| . \tag{8.6}$$

$C_R = 4/3$ or 3 is the quadratic Casimir of the quark or gluon; the transverse coordinate of the hard parton is denoted by $z = \left(x_0 + \tau\cos\phi, y_0 + \tau\sin\phi; \tau \right)$; E is the energy of the hard parton in the lab frame; $\mathbf{k}_\perp$ ($|\mathbf{k}_\perp| \le x_E E \cdot \Gamma(z)$), and $\mathbf{q}_\perp$ ($|\mathbf{q}_\perp| \le 6T(z)E \cdot \Gamma(z)$) are the local transverse momentum of the radiated gluon and the local transverse momentum transfer, respectively. The gluon fractional energy x_E and fractional plus-momentum x_+ are connected by $x_+(x_E) = x_E[1 + \sqrt{1 - (k_\perp/x_E E)^2}]/2$. We note that in the temperature range $T \sim T_c$, the coupling α_s becomes non-perturbative [27, 30, 32, 33]. Analysis of lattice data [30] suggests the following thermal running coupling form:

$$\alpha_s(Q^2) = \frac{\alpha_c}{1 + \frac{9\alpha_c}{4\pi} \log\left(\frac{Q^2}{T_c^2} \right)} , \tag{8.7}$$

with $T_c = 160$ MeV. Note that at large Q^2, Eq. (8.7) converges to vacuum running $\alpha_s(Q^2) = \frac{4\pi}{9\log(Q^2/\Lambda^2)}$; while at $Q = T_c$, $\alpha_s(T_c^2) = \alpha_c$.

The particle number density $\rho(z)$ is determined by the medium temperature $T(z)$ via $\rho(T) = \xi_s s(T)$, where $\xi_s = 0.253$ for a $N_c = 3$, $N_f = 2.5$ Stefan–Boltzmann gas, and $s(T)$ is the bulk entropy density. In the presence of hydrodynamical 4-velocity fields $u_f^\mu(z)$, boosting back to the lab frame, one should take into account a relativistic correction $\Gamma(z) = u_f^\mu n_\mu$ [34, 35], where the flow 4-velocity $u_f^\mu = \gamma_f(1, \boldsymbol{\beta}_f)$ and null hard parton 4-velocity $n^\mu = (1, \boldsymbol{\beta}_j)$. The bulk evolution profiles $(T(z), \rho(z), u_f^\mu(z))$ are generated from the VISH2+1 code [36–38] with MC-Glauber initial condition, $\tau_0 = 0.6$ fm/c, s95p-PCE equation of state (EOS), $\eta/s = 0.08$, and Cooper-Frye freeze-out temperature 120 MeV [39–44]. Event-averaged smooth profiles are embedded, and the path integrations $\int d\tau$ for jets initially produced at transverse coordinates $(\mathbf{x}_0, \phi)$ are cut off at dynamical $T(z(\mathbf{x}_0, \phi, \tau))|_{\tau_{\max}} \equiv T_{cut} = 160$ MeV hypersurfaces [9].

The Debye screening mass $\mu(z)$ is determined from solving the self-consistent equation

$$\mu^2(z) = \sqrt{4\pi\alpha_s(\mu^2(z))} T(z)\sqrt{1 + N_f/6} \tag{8.8}$$

as in [45]; $\chi^2(z) = M^2 x_+^2 + m_g^2(z)(1 - x_+)$ regulates the soft collinear divergences in the color antennae and controls the Landau–Pomeranchuk–Migdal (LPM) phase, the gluon plasmon mass $m_g(z) = f_E \mu(z)/\sqrt{2}$.

Since the sQGMP contains both chromo electrically charged quasi-particles (cec) and chromo magnetically charged quasi-particles (cmc), when jets propagate through the medium near T_c, the total quasi-particle number density ρ is divided into EQPs with fraction $\chi_T = \rho_E/\rho$ and MQPs with fraction $1 - \chi_T = \rho_M/\rho$. The parameter f_E and f_M is defined via $f_E \equiv \mu_E/\mu$ and $f_M \equiv \mu_M/\mu$, with μ_E and μ_M being the electric and magnetic screening mass, respectively, following

$$f_E(T(z))) = \sqrt{\chi_T(T(z))}\,, \qquad f_M(T(z)) = c_m\, g(T(z))\,, \qquad (8.9)$$

with the local electric "coupling" $g(T(z)) = \sqrt{4\pi\alpha_s(\mu^2(T(z)))}$.

In current sQGMP modeling, the cec component fraction χ_T remains a theoretical uncertainty related to the question of how fast the color degrees of freedom get liberated. To estimate χ_T, one notices that: (1) when temperature is high, χ_T should reach unity, i.e. $\chi_T(T \gg T_c) \to 1$; (2) in the vicinity of the regime $T \sim (1 - 3)T_c$, the renormalized expectation value of the Polyakov loop L (let us redefine $L \equiv \ell = \langle tr\mathcal{P}\exp\{ig\int_0^{1/T} d\tau A_0\}\rangle/N_c$) deviates significantly from unity, implying the suppression $\sim L$ for quarks and $\sim L^2$ for gluons in the semi-QGP model [46–49]. Consequently, in the liberation scheme (χ_T^L-scheme), we define the cec component fraction as

$$\chi_T(T) \equiv \chi_T^L(T) = c_q L(T) + c_g L^2(T) \qquad (8.10)$$

for the respective fraction of quarks and gluons, where we take the Stefan–Boltzmann (SB) fraction coefficients, $c_q = (10.5N_f)/(10.5N_f + 16)$ and $c_g = 16/(10.5N_f + 16)$, and the temperature dependent Polyakov loop $L(T)$ parameterized as (T in GeV)

$$L(T) = \left[\frac{1}{2} + \frac{1}{2}\text{Tanh}[7.69(T - 0.0726)]\right]^{10}\,, \qquad (8.11)$$

adequately fitting both the HotQCD [50] and Wuppertal-Budapest [51] Collaboration results.

On the other hand, another useful measure of the non-perturbative suppression of the color electric DOF is provided by the quark number susceptibilities [52–55]. The diagonal susceptibility is proposed as part of the order parameter for chiral symmetry breaking/restoration in [52], and plays a similar role as properly renormalized L for quark DOFs. In this scheme, we parameterize the lattice diagonal susceptibility of u quark number density, renormalized the susceptibility by its value at $T \to \infty$, as (T in GeV)

$$\tilde{\chi}_2^u(T) \equiv \frac{\chi_2^u(T)}{0.91} = \left[\frac{1}{2}\{1 + \text{Tanh}[15.65(T - 0.0607)]\}\right]^{10}\,, \qquad (8.12)$$

and define the cec component fraction in deconfinement scheme (χ_T^u-scheme) as:

$$\chi_T(T) \equiv \chi_T^u(T) = c_q \tilde{\chi}_2^u(T) + c_g L^2(T). \tag{8.13}$$

These two different schemes, for the rate of "quark liberation," with χ_T^L the "slow" and χ_T^u the "fast," provide useful estimates of theoretical systematic uncertainties associated with the quark component of the sQGMP model.

8.2 CIBJET Framework

Event-by-event fluctuations strongly influence soft flow observables. Odd harmonics are entirely determined by fluctuations. The transfer of the pattern of soft azimuthal flow fluctuations onto the pattern of hard azimuthal fluctuations requires a combined simultaneous quantitative description of both soft long wavelength and hard short wavelength dynamics. In [56, 57] the first successful simultaneous account of Soft-Hard observables R_{AA}, $v_2(p_T)$ and $v_3(p_T)$ was demonstrated using the ebe-vUSPHydro+BBMG framework. In that framework a parametric BBMG energy loss model with linear path-length dependence was used that could however not be further exploited to constrain the color composition of the QCD fluid.

Aiming to study the influence of event-by-event medium background on jet observables, we further perform jet quenching calculations on top of bulk profile simulated on an event-by-event basis, by using the viscous hydrodynamic simulation code VISHNU [44] which has been widely used and well vetted at both RHIC and the LHC. Referred to as CIBJET, the new model solves that problem by combining iEBE-VISHNU+CUJET3.1 models and is the first combined soft+hard framework with sufficient generality to test different color composition models.

8.2.1 Soft-Hard Event Engineering

Besides the fluctuating bulk background, one key difference in CIBJET framework is how azimuthal anisotropies v_n is estimated, which is experimentally measured with respect to Event Plane of corresponding order Ψ_n. In CUJET3 framework, which is based on event-averaged bulk background, the Event Plane is fixed by the Reaction Plane, i.e. ψ_n at any order is always along the direction of impact parameter **b**. Hence, we expect vanishing odd-ordered anisotropies v_{2n+1}. On the other hand, nonzero v_{2n+1} can be induced via event-by-event simulations. For instance, the fluctuation of the bulk background allows nonzero triangular geometrical component. Hence we do expect nonzero v_3 for both soft and hard sectors.

In experiment analysis, v_n's are measured via 2-particle correlations. For each collision event, one can define angular correlations of all particles as well as that of particle of interest (POI), i.e. particles in the corresponding p_T bin:

$$\langle 2 \rangle_n \equiv \langle e^{i n \phi_i} e^{-i n \phi_j} \rangle_{i \neq j}, \quad i,j \in \{\text{all charged particles}\} \tag{8.14}$$

$$\langle 2' \rangle_{n; p_T} \equiv \langle e^{i n \phi_p} e^{-i n \phi_j} \rangle_{p \neq j}, \quad j \in \{\text{all charged particles}\}, \quad p \in \{p_T \text{ bin}\} \tag{8.15}$$

and eventually

$$v_n\{2\}(p_T) = \frac{\langle \langle 2' \rangle_{n; p_T} \rangle_{\text{all events}}}{\sqrt{\langle \langle 2 \rangle_n \rangle_{\text{all events}}}}. \tag{8.16}$$

Noting that soft particles dominate in the final state, hence one can constrain the reference particles to be soft particles, and simplify the expression of correlations:

$$\langle 2 \rangle_n = (v_n^{\text{soft}})^2 \tag{8.17}$$

$$\langle 2' \rangle_{n; p_T} = \langle e^{i n (\phi_p - \psi_n^{\text{soft}})} \rangle_p \langle e^{-i n (\phi_j - \psi_n^{\text{soft}})} \rangle_j = v_n^{\text{soft}} \langle \cos[n \phi^{\text{hard}} - n \psi_n^{\text{soft}}] \rangle \tag{8.18}$$

Based on above analysis, we compute the elliptic flow v_2 and triangular flow v_3 in CIBJET simulations by soft-hard correlations:

$$v_n(p_T) \equiv \left\langle \frac{v_n^{\text{soft}}}{\langle (v_n^{\text{soft}})^2 \rangle^{1/2}} \frac{\int_0^{2\pi} d\phi \, R_{AA}(p_T, \phi) \cos[n\phi - n\psi_n^{\text{soft}}]}{\int_0^{2\pi} d\phi \, R_{AA}(p_T, \phi)} \right\rangle, \tag{8.19}$$

where v_n^{soft} and ψ_n^{soft} are the flow and Event Plane angle, respectively, of soft particles from corresponding hydrodynamic background, and $\langle \cdots \rangle$ denotes for averaging over different collision events. It is worth mentioning that in [56, 57], v_n's are defined by introducing the hard-angle $\psi_n^{\text{hard}}(p_T)$ according to

$$\int_0^{2\pi} d\phi \, R_{AA}(p_T, \phi) \sin[n\phi - n\psi_n^{\text{hard}}(p_T)] = 0,$$

and

$$v_n^{\text{hard}}(p_T) \equiv \frac{\int_0^{2\pi} d\phi \, R_{AA}(p_T, \phi) \cos[n\phi - n\psi_n^{\text{hard}}(p_T)]}{2\pi R_{AA}(p_T)}, \tag{8.20}$$

$$v_n(p_T) \equiv \frac{\langle v_n^{\text{soft}} v_n^{\text{hard}}(p_T) \cos[n\psi^{\text{hard}}(p_T) - n\psi_n^{\text{soft}}] \rangle}{\langle (v_n^{\text{soft}})^2 \rangle^{1/2}}. \tag{8.21}$$

One can easily prove the equivalence of these two definitions.

Finally, we estimate the statistical uncertainty of event-by-event quantities by the standard deviation of the mean value:

$$\delta R_{AA}(p_T) \equiv \frac{1}{\sqrt{N_{\text{events}}}} \left\{ \left\langle [R_{AA}^i(p_T) - \langle R_{AA}(p_T) \rangle]^2 \right\rangle_{i \in \{\text{all events}\}} \right\}^{1/2} \qquad (8.22)$$

$$\delta v_n(p_T) \equiv \frac{1}{\sqrt{N_{\text{events}}}} \left\{ \left\langle [v_n^i(p_T) - \langle v_n(p_T) \rangle]^2 \right\rangle_{i \in \{\text{all events}\}} \right\}^{1/2} \qquad (8.23)$$

8.2.2   Fluctuating Initial Conditions

In the event-by-event hydrodynamic simulations, one key input is the fluctuation mechanism of the initial condition of the hot medium, i.e. the initial entropy density, which is given by geometric and collisional information. From the nucleon-nucleon scattering cross-section σ_{NN} as well as the nucleon configuration of both nuclei, one can sample which nucleons are evolved in the collisions, and obtain the *participant thickness*:

$$T_{A,B}(x, y) = \int dz \, \rho_{A,B}^{\text{part}}(x, y, z), \qquad (8.24)$$

where A and B label the projectiles. Upon this picture, the Glauber model [58] further obtains the distribution of *binary collisions* as well as *"wounded" nucleons*:

$$BC(x, y) = T_A(x, y)T_B(x, y), \qquad WN(x, y) = T_A(x, y) + T_B(x, y), \qquad (8.25)$$

and assumes that the initial entropy density $s(x, y)$ is proportional to the linear combination of them:

$$s(x, y) \propto \alpha BC(x, y) + (1 - \alpha)\frac{WN(x, y)}{2}. \qquad (8.26)$$

On the other hand, the T_RENTo model [59], abbreviating for *Reduced Thickness Event-by-event Nuclear Topology* model, takes that

$$s \propto \left(\frac{T_A^p + T_B^p}{2} \right)^{1/p}, \qquad (8.27)$$

where the dimensionless parameter p determines the "reduced thickness" lying between the minimum and maximum of T_A and T_B. In this dissertation, we take $p = 0$, and the model can be simplified as

$$s \propto \sqrt{T_A T_B}. \qquad (8.28)$$

Due to the typical eccentricity of the initial entropy densities given by these two models, different parameters, for both bulk background and jet energy loss, are

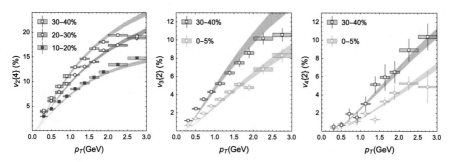

Fig. 8.1 Flow harmonics v_2, v_3, v_4 for low-p_T particles given by Event-by-Event *Monte Carlo Glauber* initial conditions, in comparison with ALICE data [60]. The width of the bands represents the statistic uncertainty for $\sim$300 events

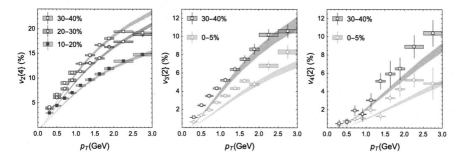

Fig. 8.2 Flow harmonics v_2, v_3, v_4 for low-p_T particles given by Event-by-Event *Monte Carlo $T_R ENTo$* initial conditions, in comparison with ALICE data [60]. The width of the bands represents the statistic uncertainty for $\sim$300 events

employed, which are hydrodynamic simulations with fixed by event-averaged initial conditions. For Glauber model, we used share viscosity $\eta/s = 0.10$; for T_RENTo model, we take $\eta/s = 0.20$.

With such bulk set-up, we obtain reasonable soft particle distribution. In Figs. 8.1 and 8.2, we show flow harmonics v_2, v_3, v_4 for low-p_T particles given by event-by-event (Glauber / T_RENTo) initial conditions, in comparison with ALICE data [60]. Both model give good agreement with v_2 and reasonable description of v_3 and v_4.

References

1. J. Liao, E. Shuryak, Phys. Rev. Lett. **102**, 202302 (2009). https://doi.org/10.1103/PhysRevLett.102.202302
2. X. Zhang, J. Liao, Phys. Rev. **C87**, 044910 (2013). https://doi.org/10.1103/PhysRevC.87.044910
3. X. Zhang, J. Liao, Phys. Rev. **C89**(1), 014907 (2014). https://doi.org/10.1103/PhysRevC.89.014907

4. S.K. Das, F. Scardina, S. Plumari, V. Greco, Phys. Lett. **B747**, 260 (2015). https://doi.org/10. 1016/j.physletb.2015.06.003

5. W.A. Horowitz, M. Gyulassy, Nucl. Phys. **A872**, 265 (2011). https://doi.org/10.1016/j. nuclphysa.2011.09.018

6. B. Betz, M. Gyulassy, Phys. Rev. **C86**, 024903 (2012). https://doi.org/10.1103/PhysRevC.86. 024903

7. K.M. Burke et al., Phys. Rev. **C90**(1), 014909 (2014). https://doi.org/10.1103/PhysRevC.90. 014909

8. A. Buzzatti, M. Gyulassy, Phys. Rev. Lett. **108**, 022301 (2012). https://doi.org/10.1103/ PhysRevLett.108.022301

9. J. Xu, A. Buzzatti, M. Gyulassy, J. High Energy Phys. **08**, 063 (2014). https://doi.org/10.1007/ JHEP08(2014)063

10. M. Cacciari, P. Nason, R. Vogt, Phys. Rev. Lett. **95**, 122001 (2005). https://doi.org/10.1103/ PhysRevLett.95.122001

11. B.A. Kniehl, G. Kramer, B. Potter, Nucl. Phys. **B582**, 514 (2000). https://doi.org/10.1016/ S0550-3213(00)00303-5

12. C. Peterson, D. Schlatter, I. Schmitt, P.M. Zerwas, Phys. Rev. **D27**, 105 (1983). https://doi. org/10.1103/PhysRevD.27.105

13. J. Xu, J. Liao, M. Gyulassy, Chin. Phys. Lett. **32**(9), 092501 (2015). https://doi.org/10.1088/ 0256-307X/32/9/092501

14. J. Xu, J. Liao, M. Gyulassy, J. High Energy Phys. **02**, 169 (2016). https://doi.org/10.1007/ JHEP02(2016)169; A.M. Polyakov, JETP Lett. **20**, 194 (1974). [,300(1974)]; B.G. Zakharov, JETP Lett. **88**, 781 (2008). https://doi.org/10.1134/S0021364008240016

15. JET Collaboration, DOE-DUKE-5396-1 (2015). https://doi.org/10.2172/1242882; B.G. Zakharov, JETP Lett. **101**(9), 587 (2015). https://doi.org/10.1134/S0021364015090131. [Pisma Zh. Eksp. Teor. Fiz.101,no.9,659(2015)]

16. N. Armesto et al., Phys. Rev. **C86**, 064904 (2012). https://doi.org/10.1103/PhysRevC.86. 064904

17. Y.T. Chien, A. Emerman, Z.B. Kang, G. Ovanesyan, I. Vitev, Phys. Rev. **D93**(7), 074030 (2016). https://doi.org/10.1103/PhysRevD.93.074030

18. E. Bianchi, J. Elledge, A. Kumar, A. Majumder, G.Y. Qin, C. Shen (2017). arXiv:1702.00481

19. S. Cao, T. Luo, G.Y. Qin, X.N. Wang, Phys. Lett. **B777**, 255 (2018). https://doi.org/10.1016/ j.physletb.2017.12.023

20. M.H. Thoma, M. Gyulassy, Nucl. Phys. **B351**, 491 (1991). https://doi.org/10.1016/S0550-3213(05)80031-8

21. J.D. Bjorken, FERMILAB-PUB-82-059-THY (1982)

22. S. Peigne, A. Peshier, Phys. Rev. **D77**, 114017 (2008). https://doi.org/10.1103/PhysRevD.77. 114017

23. M. Gyulassy, X.n. Wang, Nucl. Phys. **B420**, 583 (1994). https://doi.org/10.1016/0550-3213(94)90079-5

24. M. Gyulassy, P. Levai, I. Vitev, Nucl. Phys. **B594**, 371 (2001). https://doi.org/10.1016/S0550-3213(00)00652-0

25. M. Djordjevic, M. Gyulassy, Nucl. Phys. **A733**, 265 (2004). https://doi.org/10.1016/j. nuclphysa.2003.12.020

26. M. Djordjevic, U.W. Heinz, Phys. Rev. Lett. **101**, 022302 (2008). https://doi.org/10.1103/ PhysRevLett.101.022302

27. J. Liao, E. Shuryak, Phys. Rev. **C75**, 054907 (2007). https://doi.org/10.1103/PhysRevC.75. 054907

28. J. Liao, E. Shuryak, Phys. Rev. **C77**, 064905 (2008). https://doi.org/10.1103/PhysRevC.77. 064905

29. J. Liao, E. Shuryak, Phys. Rev. **D82**, 094007 (2010). https://doi.org/10.1103/PhysRevD.82. 094007

30. J. Liao, E. Shuryak, Phys. Rev. Lett. **101**, 162302 (2008). https://doi.org/10.1103/ PhysRevLett.101.162302

31. J. Liao, E. Shuryak, Phys. Rev. Lett. **109**, 152001 (2012). https://doi.org/10.1103/PhysRevLett.109.152001
32. B.G. Zakharov, JETP Lett. **88**, 781 (2008). https://doi.org/10.1134/S0021364008240016
33. L. Randall, R. Rattazzi, E.V. Shuryak, Phys. Rev. **D59**, 035005 (1999). https://doi.org/10.1103/PhysRevD.59.035005
34. H. Liu, K. Rajagopal, U.A. Wiedemann, J. High Energy Phys. **03**, 066 (2007). https://doi.org/10.1088/1126-6708/2007/03/066
35. R. Baier, A.H. Mueller, D. Schiff, Phys. Lett. **B649**, 147 (2007). https://doi.org/10.1016/j.physletb.2007.03.048
36. H. Song, U.W. Heinz, Phys. Rev. **C78**, 024902 (2008). https://doi.org/10.1103/PhysRevC.78.024902
37. C. Shen, U. Heinz, P. Huovinen, H. Song, Phys. Rev. **C82**, 054904 (2010). https://doi.org/10.1103/PhysRevC.82.054904
38. T. Renk, H. Holopainen, U. Heinz, C. Shen, Phys. Rev. **C83**, 014910 (2011). https://doi.org/10.1103/PhysRevC.83.014910
39. H. Song, S.A. Bass, U. Heinz, T. Hirano, C. Shen, Phys. Rev. Lett. **106**, 192301 (2011). https://doi.org/10.1103/PhysRevLett.106.192301, https://doi.org/10.1103/PhysRevLett.109.139904. [Erratum: Phys. Rev. Lett.109,139904(2012)]
40. A. Majumder, C. Shen, Phys. Rev. Lett. **109**, 202301 (2012). https://doi.org/10.1103/PhysRevLett.109.202301
41. Z. Qiu, C. Shen, U. Heinz, Phys. Lett. **B707**, 151 (2012). https://doi.org/10.1016/j.physletb.2011.12.041
42. C. Shen, U. Heinz, P. Huovinen, H. Song, Phys. Rev. **C84**, 044903 (2011). https://doi.org/10.1103/PhysRevC.84.044903
43. C. Shen, U. Heinz, Phys. Rev. **C85**, 054902 (2012). https://doi.org/10.1103/PhysRevC.86.049903, https://doi.org/10.1103/PhysRevC.85.054902. [Erratum: Phys. Rev.C86,049903(2012)]
44. C. Shen, Z. Qiu, H. Song, J. Bernhard, S. Bass, U. Heinz, Comput. Phys. Commun. **199**, 61 (2016). https://doi.org/10.1016/j.cpc.2015.08.039
45. A. Peshier (2006). arXiv:hep-ph/0601119
46. Y. Hidaka, R.D. Pisarski, Phys. Rev. **D78**, 071501 (2008). https://doi.org/10.1103/PhysRevD.78.071501
47. Y. Hidaka, R.D. Pisarski, Phys. Rev. **D81**, 076002 (2010). https://doi.org/10.1103/PhysRevD.81.076002
48. A. Dumitru, Y. Guo, Y. Hidaka, C.P.K. Altes, R.D. Pisarski, Phys. Rev. **D83**, 034022 (2011). https://doi.org/10.1103/PhysRevD.83.034022
49. S. Lin, R.D. Pisarski, V.V. Skokov, Phys. Lett. **B730**, 236 (2014). https://doi.org/10.1016/j.physletb.2014.01.043
50. A. Bazavov et al., Phys. Rev. **D80**, 014504 (2009). https://doi.org/10.1103/PhysRevD.80.014504
51. S. Borsanyi, Z. Fodor, C. Hoelbling, S.D. Katz, S. Krieg, C. Ratti, K.K. Szabo, J. High Energy Phys. **09**, 073 (2010). https://doi.org/10.1007/JHEP09(2010)073
52. L.D. McLerran, Phys. Rev. **D36**, 3291 (1987). https://doi.org/10.1103/PhysRevD.36.3291
53. S.A. Gottlieb, W. Liu, D. Toussaint, R.L. Renken, R.L. Sugar, Phys. Rev. **D38**, 2888 (1988). https://doi.org/10.1103/PhysRevD.38.2888
54. R.V. Gavai, J. Potvin, S. Sanielevici, Phys. Rev. **D40**, 2743 (1989). https://doi.org/10.1103/PhysRevD.40.2743
55. S.A. Gottlieb, W. Liu, D. Toussaint, R.L. Renken, R.L. Sugar, Phys. Rev. Lett. **59**, 2247 (1987). https://doi.org/10.1103/PhysRevLett.59.2247
56. J. Noronha-Hostler, B. Betz, J. Noronha, M. Gyulassy, Phys. Rev. Lett. **116**(25), 252301 (2016). https://doi.org/10.1103/PhysRevLett.116.252301
57. B. Betz, M. Gyulassy, M. Luzum, J. Noronha, J. Noronha-Hostler, I. Portillo, C. Ratti, Phys. Rev. **C95**(4), 044901 (2017). https://doi.org/10.1103/PhysRevC.95.044901

58. M.L. Miller, K. Reygers, S.J. Sanders, P. Steinberg, Ann. Rev. Nucl. Part. Sci. **57**, 205 (2007). https://doi.org/10.1146/annurev.nucl.57.090506.123020
59. J.S. Moreland, J.E. Bernhard, S.A. Bass, Phys. Rev. **C92**(1), 011901 (2015). https://doi.org/10.1103/PhysRevC.92.011901
60. J. Adam et al., Phys. Rev. Lett. **116**(13), 132302 (2016). https://doi.org/10.1103/PhysRevLett.116.132302

Chapter 9
Probing the Chromo-Magnetic-Monopole with Jets

With the frameworks set-up, now we are ready to test CUJET2/CUJET3 energy loss mechanism with the jet quenching observables, nuclear modification factor $R_A A$ and azimuthal anisotropy v_2, based on jet quenching on top of bulk background assuming event-averaged smooth geometries. Again, let us emphasize that the most nontrivial aspect of the CUJET3 is the chromo structure of the QGP medium when approaching $T_c \sim \Lambda_{QCD}$, which integrates two key features arising from non-perturbative dynamics pertaining to the confinement transition. The first is the suppression of chromo-electric degrees of freedom from high T toward T_c, as proposed and studied in the so-called semi-QGP model [1–3]. The second is the emergence of the chromo-magnetic degrees of freedom, i.e. the magnetic monopoles, which become dominant in the near-T_c regime and eventually reach condensation to enforce confinement at $T < T_c$, known as the "magnetic scenario" and studied extensively [4–9].

In this chapter we aim to answer this question: which scenario of decomposition is more flavored phenomenologically, the traditional quark–gluon plasma including only chromo-electric components, or the semi-quark–gluon-monopole plasma scenario with chromo-magnetic-monopoles? To this aim, we perform a series of quantitative analysis in several steps. We first perform the model parameter optimization in Sect. 9.1, based on the quantitative χ^2 analysis with a comprehensive set of experimental data for light hadrons. Then in Sect. 9.2 we show the successful CUJET3 description of available experimental data. The temperature dependence of jet transport coefficient and the corresponding shear viscosity for the quark–gluon plasma, extracted from CUJET3, are presented in Sect. 9.3. Finally the CUJET3 predictions for ongoing experimental analysis are shown in Sect. 9.4.

© Springer Nature Switzerland AG 2019
S. Shi, *Soft and Hard Probes of QCD Topological Structures in Relativistic Heavy-Ion Collisions*, Springer Theses, https://doi.org/10.1007/978-3-030-25482-7_9

9.1 CUJET Parameter Calibration

As discussed in the previous section, in the CUJET3 framework, there are two key model parameters. One parameter is α_c in Eq. (8.7), which is the value of QCD running coupling at the non-perturbative scale $Q^2 = T_c^2$ and sensitively controls the overall opaqueness of the hot medium. The other is c_m in Eq. (8.9), which is the coefficient for magnetic screening mass in the medium and influences the contribution of the magnetic component to the jet energy loss.

To systematically constrain these two key parameters, a first step we take is to perform a quantitative χ^2 analysis and utilize central and semi-central high transverse momentum light hadron's R_{AA} and v_2 for all available data. We compare the relative variance between theoretical expectation and experimental data, which is defined as the ratio of squared difference between experimental data point and corresponding CUJET3 expectation, to the quadratic sum of experimental statistic and systematic uncertainties for that data point:

$$\chi^2/\text{d.o.f.} = \sum_i \frac{(y_{\text{exp},i} - y_{\text{theo},i})^2}{\sum_s (\sigma_{s,i})^2} \bigg/ \sum_i 1, \qquad (9.1)$$

where $\sum_i$ runs over all experimental data point in the momentum range $8 \le p_T \le 50 \text{ GeV/c}$, and $\sum_s$ denotes summing over all sources of uncertainties, e.g. systematic and statistic uncertainties. We compute $\chi^2/\text{d.o.f.}$ for each of the following 12 data sets:

- 200 GeV Au–Au Collisions, 0–10% Centrality Bin, $R_{AA}(\pi^0)$: PHENIX [10, 11];
- 200 GeV Au–Au Collisions, 0–10% Centrality Bin, $v_2(\pi^0)$: PHENIX [11];
- 200 GeV Au–Au Collisions, 20–30% Centrality Bin, $R_{AA}(\pi^0)$: PHENIX [10, 11];
- 200 GeV Au–Au Collisions, 20–30% Centrality Bin, $v_2(\pi^0)$: PHENIX [11];
- 2.76 TeV Pb–Pb Collisions, 0–10% Centrality Bin, $R_{AA}(h^\pm)$: ALICE [12];
- 2.76 TeV Pb–Pb Collisions, 0–10% Centrality Bin, $v_2(h^\pm)$: ATLAS [13], CMS [14];
- 2.76 TeV Pb–Pb Collisions, 20–30% Centrality Bin, $R_{AA}(h^\pm)$: ALICE [12];
- 2.76 TeV Pb–Pb Collisions, 20–30% Centrality Bin, $v_2(h^\pm)$: ALICE [15], ATLAS [13], CMS [14];
- 5.02 TeV Pb–Pb Collisions, 0–5% Centrality Bin, $R_{AA}(h^\pm)$: ATLAS-preliminary [16], CMS [17];
- 5.02 TeV Pb–Pb Collisions, 0–5% Centrality Bin, $v_2(h^\pm)$: CMS [18];
- 5.02 TeV Pb–Pb Collisions, 10–30% Centrality Bin, $R_{AA}(h^\pm)$: CMS [17];
- 5.02 TeV Pb–Pb Collisions, 20–30% Centrality Bin, $v_2(h^\pm)$: CMS [18];

and finally obtain the overall $\chi^2/\text{d.o.f.}$ as the average over these data sets.

First of all, we perform the analysis in "slow" quark liberation scheme (χ^L_T-scheme) for a wide range of parameter space: $0.5 \le \alpha_c \le 1.3$, $0.18 \le c_m \le 0.32$. As shown in Fig. 9.1, $\chi^2/\text{d.o.f.}$ with only R_{AA} data (left panel) or only v_2 data

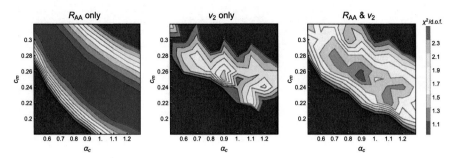

Fig. 9.1 χ^2/d.o.f. comparing χ_T^L-scheme CUJET3 results with RHIC and LHC data. Left: χ^2/d.o.f. for R_{AA} only. Middle: χ^2/d.o.f. for v_2 only. Right: χ^2/d.o.f. including both R_{AA} and v_2

(middle panel) gives different tension and favors different regions of parameter space. Taking all data together (right panel), we identify a data-selected optimal parameter set as ($\alpha_c = 0.9$, $c_m = 0.25$), with χ^2/d.o.f. close to 1, while the "uncertainty region" spanned by ($\alpha_c = 0.8$, $c_m = 0.22$) and ($\alpha_c = 1.0$, $c_m = 0.28$) with χ^2/d.o.f. about two times of the minimal value.

In order to test the necessity of chromo-magnetic-monopole degree of freedom as well as to explore potential influence of the theoretical uncertainties of different quark liberation schemes, we perform the same χ^2 analysis with two other schemes: (a) the "fast" quark liberation scheme (χ_T^u-scheme); (b) the weakly coupling QGP(wQGP) scheme, being equivalent to CUJET2.0 mode, assuming no chromo-magnetic-monopole, i.e. taking $f_E = 1$, $f_M = 0$, and cec fraction $\chi_T = 1$, while the coupling (with $\Lambda_{QCD} = 200$ MeV)

$$\alpha_s(Q^2) = \begin{cases} \alpha_{max} & \text{if } Q \leq Q_{min}, \\ \dfrac{4\pi}{9\log(Q^2/\Lambda_{QCD}^2)} & \text{if } Q > Q_{min}. \end{cases} \quad (9.2)$$

By using these three schemes with their corresponding most optimal parameter set:

(1) sQGMP χ_T^L-scheme: $\alpha_c = 0.9$, $c_m = 0.25$,
(2) sQGMP χ_T^u-scheme: $\alpha_c = 0.9$, $c_m = 0.34$,
(3) wQGP/CUJET2 scheme: $\alpha_{max} = 0.4$, (optimized by R_{AA})

we show in Fig. 9.2 their comparison with above experimental data sets, including quantitative value of χ^2/d.o.f. for each data set. While both sQGMP schemes (χ_T^L and χ_T^u) give similar jet quenching variables, the QGP scheme gives similar R_{AA} but less azimuthal anisotropy. Especially, one can see clearly from the quantitative value of their χ^2/d.o.f. that the theoretical expectations of both sQGMP schemes are in good consistency with data, and that of the QGP scheme, without cmm degree of freedom, differs significantly from the highly precise LHC v_2 measurements. The

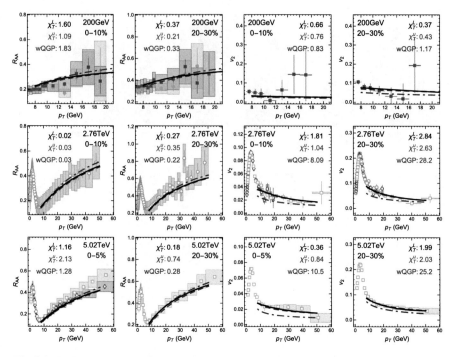

Fig. 9.2 CUJET expectation of light hadron R_{AA} and v_2 by using three different schemes: sQGMP χ_T^L-scheme (black solid), sQGMP χ_T^u-scheme (red dashed), wQGP/CUJET2 scheme (blue dashed dotted). Corresponding χ^2/d.o.f. are shown, with respect to following experimental data: PHENIX 2008 (orange solid circle) [10], PHENIX 2012 (magenta solid square) [11]; ALICE (magenta open diamond) [12, 15], ATLAS (green open circle) [13, 16], CMS (orange open square) [14, 17, 18]

χ^2 analysis strongly supports the necessity of chromo-magnetic-monopole degree of freedom, but remains robust on the specific quark liberation scheme.

While we will maintain the unification of CUJET3 model by using the same (global optimized) parameter set, it is worth mentioning that quantitative χ^2 analysis for different data set, e.g. different observable or different beam energy, flavors different parameter regime, as shown in Table 9.1. When comparing to the R_{AA} results, the azimuthal anisotropy measurement with more shrink uncertainties yields higher χ^2/d.o.f and hence stronger constraints on parameter. On the other hand, in CUJET3 models, the RHIC results flavor stronger coupling (larger α_c or c_m) than the LHC results; while the latter are more precise and give better distinction on different models. Especially with 5.02 TeV data, one can see explicitly that the sQGMP schemes are more phenomenologically flavored than the wQGP scheme.

Table 9.1 Optimal parameter and corresponding $\chi^2/d.o.f.$ for different data sets in different schemes

	sQGMP χ_T^L			sQGMP χ_T^u		wQGP	
	α_c	c_m	$\chi^2/d.o.f$	c_m	$\chi^2/d.o.f$	α_{max}	$\chi^2/d.o.f$
R_{AA}	0.9	0.24	0.57	0.31	0.60	0.4	0.67
v_2	0.9	0.25	1.34	0.34	1.28	1.0	2.34
200 GeV	1.2	0.28	0.40	0.40	0.42	0.6	0.61
2.76 TeV	0.9	0.24	1.15	0.34	1.01	1.0	2.07
5.02 TeV	0.7	0.28	0.76	0.34	1.43	1.0	8.61
All	0.9	0.25	0.97	0.34	1.02	0.7	3.47

Note that the sQGMP χ_T^u scheme is optimized with taking $\alpha_c \equiv 0.9$

9.2 Comparison with Experimental Data

With the systematic χ^2 analysis, we obtained the optimal region of CUJET3 parameters constrained by only light hadron R_{AA} and v_2, for central and semi-central collisions. To provide a critical independent test of the model, we compute CUJET3 results for both light and heavy flavor hadrons, with all centrality ranges up to semi-peripheral collisions, and perform apple-to-apple comparisons with all available experimental data.

Starting from this section, in CUJET3 simulations we employed the χ_T^L-scheme assuming slow quark-liberation, with keeping the theoretical uncertainties by taking the parameter region spanned by $(\alpha_c = 0.8, c_m = 0.22)$ and $(\alpha_c = 1.0, c_m = 0.28)$, which correspond to upper/lower bounds of R_{AA} and lower/upper bounds of v_2, respectively.

9.2.1 Light Hadrons

First of all, in Figs. 9.3, 9.4, 9.5, 9.6, 9.7, we compare CUJET3 results for light hadrons' R_{AA} and v_2, with all available data: PHENIX [10, 11] and STAR [19] measurements for 200 GeV Au–Au collisions; ALICE [12, 15], ATLAS [13, 20] and CMS [14, 21] results for 2.76 TeV Pb–Pb collisions; and ATLAS [16] and CMS [17, 18] data for 5.02 TeV Pb–Pb collisions. One can clearly see the excellent agreement for all centrality range at all these collision energies.

It is worth mentioning that while varying the centrality classes from 0–5% to 40–50%, the bulk background gets cooler, smaller in size, shorter lived, and more elliptic. In such systems, the path length of the jets, either direction averaging or depending, varies in a wide range. The success in explaining R_{AA} and v_2 from central to semi-peripheral data indicates the success of path dependence of the CUJET3 energy loss model.

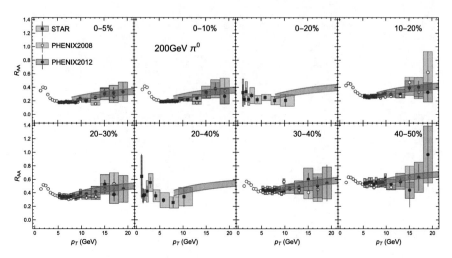

Fig. 9.3 Light hadron R_{AA} for 200 GeV Au–Au collisions in comparison with PHENIX [10, 11] and STAR [19] results

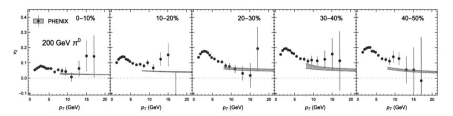

Fig. 9.4 Light hadron v_2 for 200 GeV Au–Au collisions in comparison with PHENIX data [11]

9.2.2 Heavy Flavor Measurements

With successfully describing high-p_T R_{AA} and v_2 data for light hadrons, we further test the energy loss mechanism for heavy quarks. In Figs. 9.8, 9.9, 9.10, 9.11, 9.12, 9.13, 9.14, 9.15, we compare CUJET3 results for the energy loss observables of prompt D & B mesons as well as electrons or muons from heavy flavor decay, with all available data: PHENIX [22], STAR [23] measurements for 200 GeV Au–Au collisions; ALICE [24–28], CMS [29] data for 200 GeV Pb–Pb collisions; and finally CMS results [30–32] for 5.02 TeV Pb–Pb collisions. One again observes very good agreement between model and data, validating a successful unified description of CUJET3 for both light and heavy flavor jet energy loss observables.

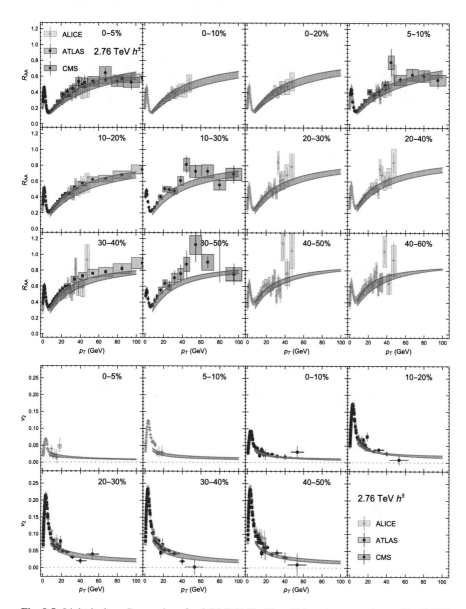

Fig. 9.5 Light hadron R_{AA} and v_2 for 2.76 TeV Pb–Pb collisions in comparison with ALICE [12, 15], ATLAS [13, 20] and CMS [14, 21] results

9.3 Jet Transport Coefficient and Shear Viscosity

As discussed above, the jet quenching observables of light hadrons provide stringent constraints on values of the jet energy loss parameters. In the meanwhile, the comparison between three different schemes, (1) sQGMP-χ_T^L, (2) sQGMP-χ_T^u, and

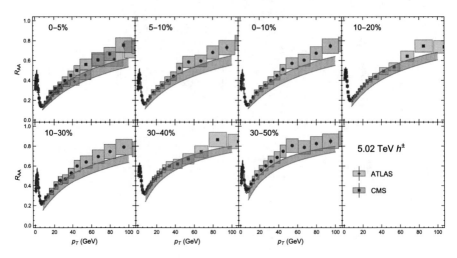

Fig. 9.6 Light hadron R_{AA} for 5.02 TeV Pb–Pb collisions in comparison with ATLAS [16] and CMS [17] results

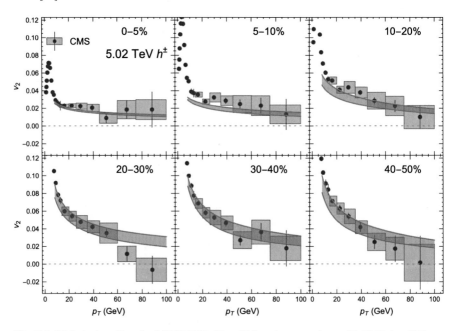

Fig. 9.7 Light hadron R_{AA} for 2.76 TeV Pb–Pb collisions in comparison with CMS data [18]

(3) wQGP, shows the necessity of chromo-magnetic-monopole degree of freedom, while robustness on quark liberation rate. It is of great interests to further compare how the jet and bulk transport properties differ in these schemes. This will pave the way for clarifying the temperature dependence of jet quenching and shear viscous transport properties based on available high p_T data in high-energy A+A collisions.

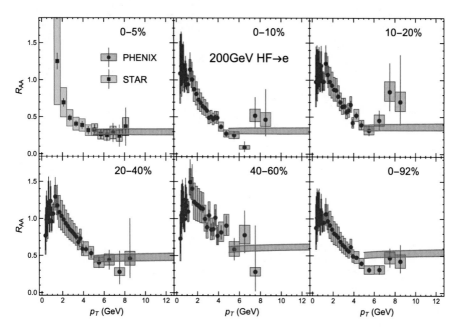

Fig. 9.8 Heavy flavor decayed electron R_{AA} for 200 GeV Au–Au collisions in comparison with PHENIX [22] and STAR [23] results

The jet transport coefficient $\hat{q}$ characterizes the averaged transverse momentum transfer squared per mean free path [33]. For a quark jet (in the fundamental representation F) with initial energy E, we calculate its $\hat{q}$ in the same way as the previous CUJET3.0 computation in [34, 35], via

$$\hat{q}_F(E, T) = \int_0^{6ET} dq_\perp^2 \frac{2\pi}{(q_\perp^2 + f_E^2 \mu^2(z))(q_\perp^2 + f_M^2 \mu^2(z))} \rho(T)$$
$$\times \left\{ \left[C_{qq} f_q + C_{qg} f_g \right] \cdot \left[\alpha_s^2(q_\perp^2) \right] \cdot \left[f_E^2 q_\perp^2 + f_E^2 f_M^2 \mu^2(z) \right] \right.$$
$$\left. + \left[C_{qm}(1 - f_q - f_g) \right] \cdot [1] \cdot \left[f_M^2 q_\perp^2 + f_E^2 f_M^2 \mu^2(z) \right] \right\}, \quad (9.3)$$

and similarly for a gluon/cmm jet:

$$\hat{q}_g(E, T) = \int_0^{6ET} dq_\perp^2 \frac{2\pi}{(q_\perp^2 + f_E^2 \mu^2(z))(q_\perp^2 + f_M^2 \mu^2(z))} \rho(T)$$
$$\times \left\{ \left[C_{gq} f_q + C_{gg} f_g \right] \left[\alpha_s^2(q_\perp^2) \right] \left[f_E^2 q_\perp^2 + f_E^2 f_M^2 \mu^2(z) \right] \right.$$
$$\left. + \left[C_{gm}(1 - f_q - f_g) \right] [1] \left[f_M^2 q_\perp^2 + f_E^2 f_M^2 \mu^2(z) \right] \right\}, \quad (9.4)$$

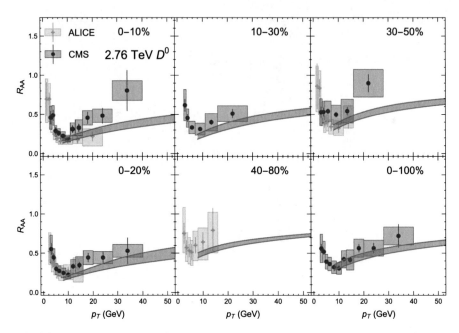

Fig. 9.9 Prompt D meson R_{AA} for 2.76 TeV Pb–Pb collisions in comparison with ALICE [24] and preliminary CMS [29] results

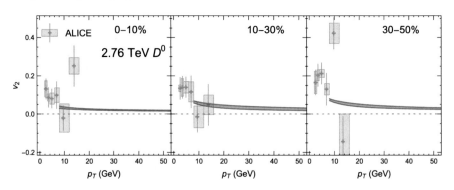

Fig. 9.10 D meson v_2 for 2.76 TeV Pb–Pb collisions in comparison with ALICE data [25]

$$\hat{q}_m(E, T) = \int_0^{6ET} dq_\perp^2 \frac{2\pi}{(\boldsymbol{q}_\perp^2 + f_E^2 \mu^2(z))(\boldsymbol{q}_\perp^2 + f_M^2 \mu^2(z))} \rho(T)$$
$$\times \left\{ \left[C_{mq} f_q + C_{mg} f_g \right] [1] \left[f_E^2 \boldsymbol{q}_\perp^2 + f_E^2 f_M^2 \mu^2(z) \right] \right.$$
$$\left. + \left[C_{mm}(1 - f_q - f_g) \right] \left[\alpha_s^{-2}(\boldsymbol{q}_\perp^2) \right] \left[f_M^2 \boldsymbol{q}_\perp^2 + f_E^2 f_M^2 \mu^2(z) \right] \right\}.$$
$$(9.5)$$

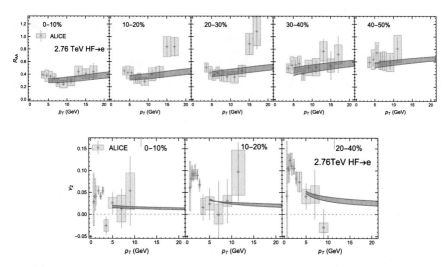

Fig. 9.11 Heavy flavor decayed electron R_{AA} and v_2 for 2.76 TeV Pb–Pb collisions in comparison with ALICE data [26, 27]

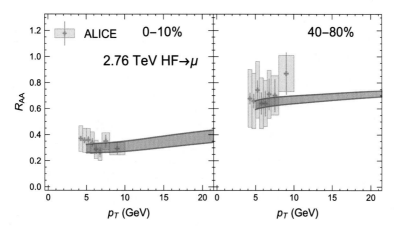

Fig. 9.12 Heavy flavor decayed muon R_{AA} for 2.76 TeV Pb–Pb collisions in comparison with ALICE data [28]

The quasi-parton density fractions of quark (q) or gluon (g), denoted as $f_{q,g}$, are defined as

$$f_q = c_q L(T), \ f_g = c_g L(T)^2, \qquad (\text{if } \chi_T^L)$$
$$f_q = c_q \tilde{\chi}_2^u(T), \ f_g = c_g L(T)^2, \qquad (\text{if } \chi_T^u) \tag{9.6}$$

respectively, for sQGMP χ_T^L and χ_T^u scheme. The magnetically charged quasi-particle density fraction is hence $f_m = 1 - \chi_T = 1 - f_q - f_g$. The color factors are given by

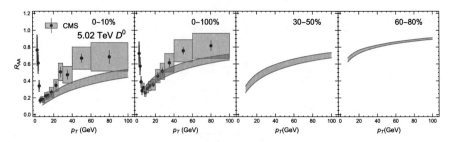

Fig. 9.13 Prompt D meson R_{AA} for 5.02 TeV Pb–Pb collisions in comparison with preliminary CMS data [30]

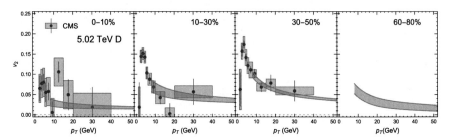

Fig. 9.14 Prompt D meson v_2 for 5.02 TeV Pb–Pb collisions in comparison with preliminary CMS data [32]

Fig. 9.15 Prompt B meson R_{AA} for 5.02 TeV Pb–Pb collisions in comparison with CMS data [31]

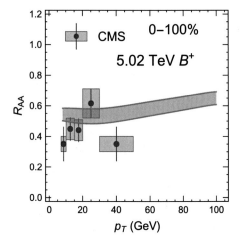

$$C_{qq} = \frac{4}{9}, \; C_{gg} = C_{mm} = C_{gm} = C_{mg} = \frac{9}{4},$$

$$C_{qg} = C_{gq} = C_{qm} = C_{mq} = 1 \, .$$
(9.7)

One can find that while switching to the wQGP scheme, by taking $f_q = c_q$, $f_g = c_g$, $f_E = 1$, $f_M = 0$, turning off the cmm channel, and employing the running

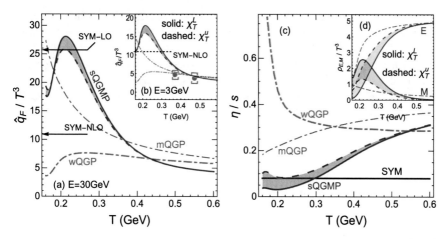

Fig. 9.16 (Left) The temperature dependence of the dimensionless jet transport coefficient $\hat{q}_F/T^3$ for a light quark jet with initial energy $E =$. (**a**) 30 GeV, (**b**) 3 GeV in the CUJET framework with the three schemes: (1) sQGMP-χ_T^L scheme (red solid), (2) sQGMP-χ_T^u scheme (red dashed), and (3) wQGP/CUJET2.0 scheme (green dotdashed). The $N = 4$ leading order/next to leading order Super Yang-Mills $\hat{q}_{SYM-LO}/T^3 = \frac{\pi^{3/2}\Gamma(\frac{3}{4})}{\Gamma(\frac{5}{4})}\sqrt{\lambda}$ and $\hat{q}_{SYM-NLO}/T^3 = \frac{\pi^{3/2}\Gamma(\frac{3}{4})}{\Gamma(\frac{5}{4})}\sqrt{\lambda}(1 - \frac{1.957}{\sqrt{\lambda}})$, respectively, [40] with coupling $\lambda = 4\pi \cdot 3 \cdot 0.31$, are plotted for comparisons. The green blobs in inset (**b**) shows the JET collaboration [33] model average of $\hat{q}_F/T^3$ while the boxes represent the uncertainties. (Right) The shear viscosity to entropy density ratio η/s estimated with scheme (1) (red solid), (2) (red dashed), and (3) (green dotdashed). The inset shows quasi-particle number density fraction of q, g, m in the liberation scheme χ_T^L (solid) and χ_T^u (dashed)

coupling $\alpha_s(Q^2)$ defined in Eq. (9.2), the jet transport coefficient $\hat{q}$ for a quark/gluon jet defined in Eq. (9.3/9.4) returns to that of the CUJET2.0 framework [36].

Once the jet transport coefficient $\hat{q}$ has been computed, one can extrapolate $\hat{q}(T, E)$ down to thermal energy scales $E \sim 3T/2$ and estimate the shear viscosity to entropy density ratio η/s, based on kinetic theory in a weakly coupled quasi-particle picture [37–39]. An estimate of η/s can be derived as

$$
\begin{aligned}
\eta/s &= \frac{1}{s}\frac{4}{15}\sum_a \rho_a \langle p \rangle_a \lambda_a^\perp \\
&= \frac{4T}{5s}\sum_a \rho_a \left(\sum_b \rho_b \int_0^{\langle S_{ab}\rangle/2} dq_\perp^2 \frac{4q_\perp^2}{\langle S_{ab}\rangle}\frac{d\sigma_{ab}}{dq_\perp^2} \right)^{-1} \\
&= \frac{18T^3}{5s}\sum_a \rho_a/\hat{q}_a(T, E = 3T/2) \ .
\end{aligned}
\tag{9.8}
$$

The $\rho_a(T) \equiv f_a \, \rho(T)$ is the quasi-parton density of type $a = q, g, m$. The mean thermal Mandelstam variable $\langle S_{ab} \rangle \sim 18T^2$. Clearly the η/s of the system is dominated by the ingredient which has the largest $\rho_a/\hat{q}_a$.

Further insights on the viability of a unified and consistent understanding of the soft and hard sectors together can be obtained by investigating the corresponding soft and hard transport properties of QGP. That is, one could try to calculate the $\hat{q}_F/T^3$ and η/s for a given QGP medium model whose parameters have been calibrated with data. Here we explore three models with distinct chromo structure: a wQGP medium (as in CUJET2) with only chromo-electric component, a mQGP medium with unsuppressed chromo-electric component plus an added magnetic component, as well as a sQGMP medium (as in CIBJET) with suppressed chromo-electric component and an emergent magnetic component. These coefficients are computed by properly synthesizing the contributions from all components to the momentum-square transfer with a jet (in the case of $\hat{q}_F/T^3$) or to the scattering cross-sections (in the case of η/s). The detailed formulae can be found in e.g. [34, 35].

The results are shown in Fig. 9.16. Both sQGMP and mQGP models show a strong enhancement of $\hat{q}_F/T^3$ and the decrease of η/s in the near T_c regime, an important feature that is absent in the wQGP and is due to the emergence of the magnetic component. Such nontrivial near-T_c features are much stronger in sQGMP than in mQGP, which could be understood from the fact that the magnetic component is more dominant in sQGMP (i.e. the ratio of magnetic density to electric density is larger, see panel (d) of Fig. 9.16). In terms of hard sector, it is already apparent that the v_2 at high p_T favors the sQGMP model. For the soft sector, the η/s comparison also clearly favors the sQGMP model, leading to $\eta/s \sim (0.1-0.2)$ in the relevant temperature regime which are precisely the needed values (for either Glauber or Trento initial conditions) for hydro calculations to correctly produce the bulk soft anisotropy observables v_2 and v_3. We note in passing that the sQGMP transport coefficients around T_c are close to the values suggested from strongly coupled field theories via AdS/CFT approach [40, 41].

Within the sQGMP scenario, we've further studied two slightly different suppression schemes for the chromo-electric component. The χ_T^L scheme uses the lattice-computed Polyakov loop as a "penalty" for color charge to characterize the suppression of quark sector as in the original semi-QGP. The χ_T^u scheme instead uses the lattice-computed quark number susceptibilities to quantify the suppression of quark sector. In both schemes the suppression of gluon sector is based on Polyakov loop as in semi-QGP. The main difference is that there is stronger (faster) suppression in the χ_T^L scheme than the χ_T^u scheme of chromo-electric component from high T toward low T: see the solid versus dashed curves in the panel (d) of Fig. 9.16. With both schemes phenomenologically viable, the χ_T^u scheme seems preferred by virtue of consistency with the KSS bound $\eta/s \geq 1/4\pi$ [41].

9.4 CUJET3 Predictions for Ongoing Experimental Analysis

In above section we perform a successful test of the CUJET3 framework, which provides a united description for comprehensive sets of experimental data, from average suppression to azimuthal anisotropy, from light flavor to heavy flavor observables, with beam energy from 200 GeV to 5.02 TeV, and from central to semi-peripheral collisions. With new colliding system or new experimental observables, we expect more stringent test to help further constrain the CUJET3 energy loss model. In this section, we show the CUJET3 prediction for ongoing experimental analysis, including jet quenching observables in $^{129}_{54}$Xe–$^{129}_{54}$Xe collisions at 5.44 TeV and more heavy flavor signals in 5.02 TeV Pb–Pb collisions.

9.4.1 Light Hadron R_{AA} in 5.44 TeV $^{129}_{54}$Xe–$^{129}_{54}$Xe Collisions

Recently the LHC ran collisions with a new species of nuclei, colliding xenon with 129 nucleons ($^{129}_{54}$Xe), with beam energy $\sqrt{s_{NN}} = 5.44$ TeV. In Xe–Xe collisions, the hot medium created is expected to be a bit cooler and shorter lived when comparing with the one created in 5.02 TeV Pb–Pb collisions. Given the similar beam energy, it is expected that the difference between observables from these two colliding system should provide valuable information on the nature of the QGP, especially on how the hot medium interacts with high energy jets.

In Fig. 9.17 we show the light hadron R_{AA} and v_2 for both systems. One can clearly see that it produces higher R_{AA} and lower v_2 in 5.44 TeV Xe–Xe collisions (blue bands), when comparing with those in 5.02 TeV Pb–Pb collisions (red dashed curves). It indicates the high-p_T light hadrons produced in the former system are less suppressed than those produced in latter. This shows the sensitivity of the jet quenching observables to the system size and density: when comparing to those created in Pb–Pb collisions, jets created in Xe–Xe collisions travel with shorter

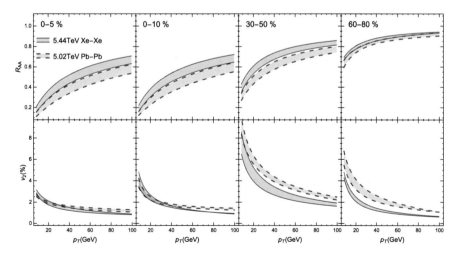

Fig. 9.17 Light hadron R_{AA} and v_2 for 5.44 TeV Xe–Xe collisions (blue bands) and 5.02 TeV Pb–Pb collisions (red dashed curves)

path in the hot medium and interact with less dense matter, hence they lose less energy. With this new colliding system, we are able to further test the path length dependence of the CUJET3 jet energy loss model.

9.4.2   B-Decayed D Meson R_{AA} in 5.02 TeV Pb–Pb Collisions

Another new experimental measurement is the B-decayed D meson R_{AA} in 5.02 TeV Pb–Pb collisions. As shown in Fig. 9.18, the R_{AA} of B-decay D meson (left panel) has similar p_T-dependence as that of B mesons (right panel), and both of them are less suppressed than the prompt D meson (middle panel), especially for the region with lower momentum ($p_T < 20$ GeV). We expect that future precise measurement of B-decay D meson R_{AA} should provide observation of the "dead cone" effect which suppresses the radiational energy loss of bottom jets.

9.4.3   High-p_T D Mesons in 200 GeV Au–Au Collisions

Recently the STAR Collaboration at RHIC installed the Heavy Flavor Tracker, which allows high precision measures of open heavy flavor hadrons. Early results of azimuthal anisotropy for lower p_T D mesons have shown interesting property of the low energy charm quarks [42]. With the CUJET3 predictions for D meson's R_{AA} and v_2 shown in Fig. 9.19, we expect that precise measurements of high p_T D mesons' jet quenching observable could allow us the direct comparison with heavy flavor data, and further test the consistency of HF sector of CUJET3 energy loss, for different beam energies.

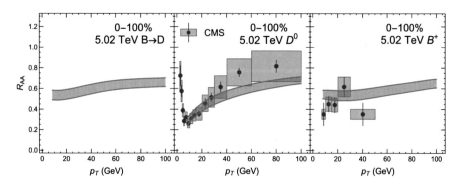

Fig. 9.18 R_{AA} for D meson from B-decay (left), prompt D meson (middle), and B meson (right) in minimal-bias 2.76 TeV Pb–Pb collisions

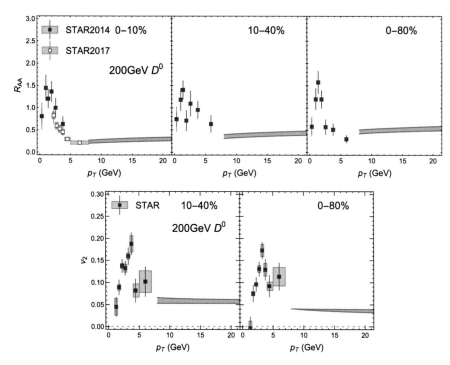

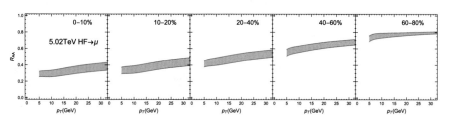

Fig. 9.19 D meson's R_{AA} and v_2 in 200 GeV Au–Au collisions. STAR data [42, 43] for lower p_T range are also shown

Fig. 9.20 R_{AA} for heavy flavor decayed muon in 5.02 TeV Pb–Pb collisions

9.4.4 Heavy Flavor Decayed Leptons in 5.02 TeV Pb–Pb Collisions

Finally we show the CUJET3 predictions for heavy flavor decayed muons and electrons in Figs. 9.20 and 9.21. Being the decay product of both D and B mesons, the R_{AA} in lower p_T regime is sensitive to relative ratios between D and B absolute cross-sections. We expect more stringent future test from the heavy flavor sector to help further constrain CUJET3.

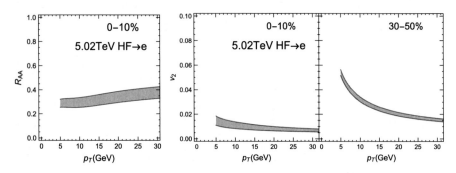

Fig. 9.21 R_{AA}(left) and v_2(right) for heavy flavor decayed muon in 5.02 TeV Pb–Pb collisions

9.5   Summary of CUJET 3.1 Results

In this chapter we performed a series of quantitative analysis on the jet quenching observables, nuclear modification factor R_{AA} and azimuthal anisotropy v_2, based on jet quenching on top of bulk background assuming event-averaged smooth geometries. By comparing experimental data and model results with different assumptions on the constituent of the hot medium, the scenario with *chromo-magnetic-monopoles* is flavored, rather than the traditional composition of simply quarks and gluons. The chromo-magnetic-monopole scenario, referred to as a semi-Quark-Gluon-Monopole Plasma (sQGMP), involves suppressed color electric quark and gluon degrees of freedom $q(T) + g(T)$ as well as the emergent color magnetic monopole degrees of freedom, $m(T)$, with their temperature dependence as implied by available lattice QCD data.

We first calibrate the two control parameters of the energy loss model: the maximum value of the running QCD coupling, α_c, and the ratio of the screening masses, $c_m = \mu_M/\mu_E$. We performed a global quantitative χ^2 analysis to utilize light hadron jet quenching observables for central and semi-central heavy-ion collisions for beam energy $\sqrt{s_{NN}} = 200\,\text{GeV(Au–Au)}$, 2.76 TeV(Pb–Pb), and 5.02 TeV(Pb–Pb). The global χ^2 is found to be minimized to near unity for $\alpha_c \approx 0.9\pm0.1$, and $c_m \approx 0.25\pm0.03$. With significantly smaller χ^2 in compare with that of CUJET2 model, such quantitative analysis strongly supports the necessity of chromo-magnetic-monopole degree of freedom.

By taking the data-selected optimal parameters, the CUJET3 framework provides a unified, systematic and successful description of a comprehensive set of available data, from average suppression to azimuthal anisotropy, from light to heavy flavors, from central to semi-peripheral collisions, for all three colliding systems. Besides the successful test of CUJET3, the semi-Quark-Gluon-Monopole Plasma could provide a consistent description of both soft and hard transport properties indicated by heavy ion data.

An important theoretical advantage of the CUJET3.1 framework is that it is not only χ^2 consistent with soft and hard observables data at RHIC and LHC, but also with non-perturbative lattice QCD data, and, most remarkably, also with a viscosity to entropy ratio ($\eta/s \sim T^3/\hat{q} \sim 0.1$) required for internal consistency of the sQGMP jet transport coefficient, $\hat{q}$, with the perfect fluidity property of QCD fluids near T_c.

We finally present CUJET3 predictions for ongoing experimental analysis. We expect that the comparison between the light hadrons' R_{AA} in 5.44 TeV Xe–Xe collisions and those observed in 5.02 TeV Pb–Pb collisions could further test the path-length dependence of the CUJET3 jet energy loss model. In the meanwhile, according to CUJET3 predictions for B-decayed D mesons' R_{AA} in 5.02 TeV Pb–Pb collisions, we expect that precise measurement of this observable should provide observation of the "dead cone" effect which suppresses the radiational energy loss of bottom jets.

References

1. Y. Hidaka, R.D. Pisarski, Phys. Rev. **D78**, 071501 (2008). https://doi.org/10.1103/PhysRevD.78.071501
2. Y. Hidaka, R.D. Pisarski, Phys. Rev. **D81**, 076002 (2010). https://doi.org/10.1103/PhysRevD.81.076002
3. S. Lin, R.D. Pisarski, V.V. Skokov, Phys. Lett. **B730**, 236 (2014). https://doi.org/10.1016/j.physletb.2014.01.043
4. J. Liao, E. Shuryak, Phys. Rev. **C75**, 054907 (2007). https://doi.org/10.1103/PhysRevC.75.054907
5. A. D'Alessandro, M. D'Elia, Nucl. Phys. **B799**, 241 (2008). https://doi.org/10.1016/j.nuclphysb.2008.03.002
6. J. Liao, E. Shuryak, Phys. Rev. **C77**, 064905 (2008). https://doi.org/10.1103/PhysRevC.77.064905
7. J. Liao, E. Shuryak, Phys. Rev. Lett. **101**, 162302 (2008). https://doi.org/10.1103/PhysRevLett.101.162302
8. C. Ratti, E. Shuryak, Phys. Rev. **D80**, 034004 (2009). https://doi.org/10.1103/PhysRevD.80.034004
9. C. Bonati, M. D'Elia, Nucl. Phys. **B877**, 233 (2013). https://doi.org/10.1016/j.nuclphysb.2013.10.004
10. A. Adare et al., Phys. Rev. Lett. **101**, 232301 (2008). https://doi.org/10.1103/PhysRevLett.101.232301
11. A. Adare et al., Phys. Rev. **C87**(3), 034911 (2013). https://doi.org/10.1103/PhysRevC.87.034911
12. B. Abelev et al., Phys. Lett. **B720**, 52 (2013). https://doi.org/10.1016/j.physletb.2013.01.051
13. G. Aad et al., Phys. Lett. **B707**, 330 (2012). https://doi.org/10.1016/j.physletb.2011.12.056
14. S. Chatrchyan et al., Phys. Rev. Lett. **109**, 022301 (2012). https://doi.org/10.1103/PhysRevLett.109.022301
15. B. Abelev et al., Phys. Lett. **B719**, 18 (2013). https://doi.org/10.1016/j.physletb.2012.12.066
16. ATLAS collaboration, ATLAS-CONF-2017-012 (2017)
17. V. Khachatryan et al., J. High Energy Phys. **04**, 039 (2017). https://doi.org/10.1007/JHEP04(2017)039
18. A.M. Sirunyan et al., Phys. Lett. **B776**, 195 (2018). https://doi.org/10.1016/j.physletb.2017.11.041

19. B.I. Abelev et al., Phys. Rev. **C80**, 044905 (2009). https://doi.org/10.1103/PhysRevC.80. 044905
20. G. Aad et al., J. High Energy Phys. **09**, 050 (2015). https://doi.org/10.1007/JHEP09(2015)050
21. S. Chatrchyan et al., Eur. Phys. J. **C72**, 1945 (2012). https://doi.org/10.1140/epjc/s10052-012-1945-x
22. A. Adare et al., Phys. Rev. **C84**, 044905 (2011). https://doi.org/10.1103/PhysRevC.84.044905
23. B.I. Abelev et al., Phys. Rev. Lett. **98**, 192301 (2007). https://doi.org/10.1103/PhysRevLett.106.159902, https://doi.org/10.1103/PhysRevLett.98.192301. [Erratum: Phys. Rev. Lett.106,159902(2011)]
24. B. Abelev et al., J. High Energy Phys. **09**, 112 (2012). https://doi.org/10.1007/JHEP09(2012)112
25. B.B. Abelev et al., Phys. Rev. **C90**(3), 034904 (2014). https://doi.org/10.1103/PhysRevC.90. 034904
26. J. Adam et al., Phys. Lett. **B771**, 467 (2017). https://doi.org/10.1016/j.physletb.2017.05.060
27. J. Adam et al., J. High Energy Phys. **09**, 028 (2016). https://doi.org/10.1007/JHEP09(2016)028
28. B. Abelev et al., Phys. Rev. Lett. **109**, 112301 (2012). https://doi.org/10.1103/PhysRevLett. 109.112301
29. CMS Collaboration, CMS-PAS-HIN-15-005 (2015)
30. A.M. Sirunyan et al., Phys. Lett. **B782**, 474 (2018). https://doi.org/10.1016/j.physletb.2018. 05.074
31. A.M. Sirunyan et al., Phys. Rev. Lett. **119**(15), 152301 (2017). https://doi.org/10.1103/PhysRevLett.119.152301
32. A.M. Sirunyan et al., Phys. Rev. Lett. **120**(20), 202301 (2018). https://doi.org/10.1103/PhysRevLett.120.202301
33. K.M. Burke et al., Phys. Rev. **C90**(1), 014909 (2014). https://doi.org/10.1103/PhysRevC.90. 014909
34. J. Xu, J. Liao, M. Gyulassy, Chin. Phys. Lett. **32**(9), 092501 (2015). https://doi.org/10.1088/0256-307X/32/9/092501
35. J. Xu, J. Liao, M. Gyulassy, J. High Energy Phys. **02**, 169 (2016). https://doi.org/10.1007/JHEP02(2016)169
36. J. Xu, A. Buzzatti, M. Gyulassy, J. High Energy Phys. **08**, 063 (2014). https://doi.org/10.1007/JHEP08(2014)063
37. P. Danielewicz, M. Gyulassy, Phys. Rev. **D31**, 53 (1985). https://doi.org/10.1103/PhysRevD. 31.53
38. T. Hirano, M. Gyulassy, Nucl. Phys. **A769**, 71 (2006). https://doi.org/10.1016/j.nuclphysa. 2006.02.005
39. A. Majumder, B. Muller, X.N. Wang, Phys. Rev. Lett. **99**, 192301 (2007). https://doi.org/10. 1103/PhysRevLett.99.192301
40. H. Liu, K. Rajagopal, U.A. Wiedemann, Phys. Rev. Lett. **97**, 182301 (2006). https://doi.org/10.1103/PhysRevLett.97.182301
41. P. Kovtun, D.T. Son, A.O. Starinets, Phys. Rev. Lett. **94**, 111601 (2005). https://doi.org/10. 1103/PhysRevLett.94.111601
42. L. Adamczyk et al., Phys. Rev. Lett. **118**(21), 212301 (2017). https://doi.org/10.1103/PhysRevLett.118.212301
43. L. Adamczyk et al., Phys. Rev. Lett. **113**(14), 142301 (2014). https://doi.org/10.1103/PhysRevLett.113.142301

Chapter 10
Jet-Quenching on Top of Fluctuating Hot Medium

Event-by-event fluctuations strongly influence soft flow observables. Odd harmonics are entirely determined by fluctuations. The transfer of the pattern of soft azimuthal flow fluctuations onto the pattern of hard azimuthal fluctuations requires a combined simultaneous quantitative description of both soft long wavelength and hard short wavelength dynamics. In [1, 2] the first successful simultaneous account of Soft-Hard observables R_{AA}, $v_2(p_T)$ and $v_3(p_T)$ was demonstrated using the ebe-vUSPHydro+BBMG framework. In that framework a parametric BBMG energy loss model with linear path-length dependence was used that could however not be further exploited to constrain the color composition of the QCD fluid. The CIBJET model solves that problem by combining iEBE-VISHNU+CUJET3.1 models and is the first combined soft+hard framework with sufficient generality to test different color composition models. The default composition option, referred to as a semi-Quark-Gluon-Monopole Plasma (sQGMP), involves suppressed color electric quark and gluon degrees of freedom $q(T) + g(T)$ as well as the emergent color magnetic monopole degrees of freedom, $m(T)$, with their temperature dependence as implied by available lattice QCD data. In [3–6] we showed that this composition accounts well for RHIC and LHC data at least in the simplified approximation when event-averaged smooth geometries are assumed.

Adopted from the CUJET3 energy model, in CIBJET framework [7] there are two key parameters in this framework: α_c which is the non-perturbative coupling strength at the transition temperature scale $T_c \simeq 160$ MeV; and c_m which controls the magnetic screening mass and sensitively influences the scattering rates involving the magnetic component. A recent comprehensive comparison of the model calculations (based on smooth-hydro background) for R_{AA} and v_2 with extensive data from RHIC to LHC [8–12] has allowed us to optimize these parameters. The α_c most sensitively controls overall opaqueness and is fixed as $\alpha_c = 0.9$; the c_m strongly influences anisotropy v_2 and is fixed as $c_m = 0.25$ for Glauber geometry while $c_m = 0.22$ for Trento geometry.

© Springer Nature Switzerland AG 2019
S. Shi, *Soft and Hard Probes of QCD Topological Structures
in Relativistic Heavy-Ion Collisions*, Springer Theses,
https://doi.org/10.1007/978-3-030-25482-7_10

10.1   A Unified Soft-Hard Description with CIBJET

With the above CIBJET setup, we've performed the highly demanding event-by-event simulations for sophisticated jet energy loss calculations at the LHC energy. It may be noted that computation for each centrality costs about 60,000 cpu hours. In each event, the bulk medium evolves from the hydro component while on top of that about 5×10^5 jet paths are sampled for energy loss calculation (as well as accounting for path fluctuations and gluon emission sampling). With such computing power, we are able to quantitatively explore both soft and hard observables in a unified simulation framework and to answer the aforementioned pressing questions.

The CIBJET results for nuclear modification factor R_{AA} as well as the second and third harmonic coefficients v_2 and v_3 of the final hadron azimuthal distribution are shown in Fig. 10.1 for 30–40% Pb+Pb collisions at 5.02 ATeV. It should be particularly noted that the anisotropy observables are computed in the same way as the experimental analysis on an event-wise basis. The solid curves are from event-by-event CIBJET with either Monte Carlo Glauber (red) initial condition and $\eta/s = 0.1$ or Trento (blue) initial condition and $\eta/s = 0.2$, while the dashed curves are single-shot calculations with the corresponding averaged smooth geometry. Observables in the soft region ($p_T \lesssim 3\,\mathrm{GeV}$) are computed from the hydrodynamic component while observables in the hard region ($p_T \gtrsim 10\,\mathrm{GeV}$) are computed from the jet energy loss component. The CIBJET results in both

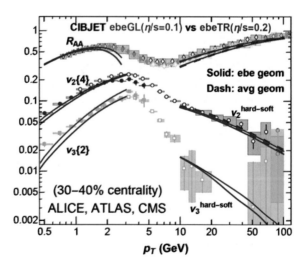

Fig. 10.1 The nuclear modification factor R_{AA} as well as the second and third harmonic coefficients v_2 and v_3 of the final hadron azimuthal distribution as functions of p_T for 30–40% Pb+Pb collisions at 5.02 ATeV. The solid curves are from event-by-event calculations while the dashed from average-geometry. The CIBJET results in both soft and hard regions, with either Monte Carlo Glauber (red) or Trento (blue) initial conditions, are in excellent agreement with experimental data from ALICE, ATLAS and CMS [8–12]

soft and hard regions, spanning a broad transverse momentum window from 0.5 to 100 GeV, are in excellent agreement with available experimental data from ALICE, ATLAS, and CMS collaborations at the LHC [8–12]. The results for the v_3 at high p_T deserve special note, which could not possibly be computed without event-by-event simulations and which serves as a further independent test of the CIBJET's phenomenological success.

One important issue is whether the high p_T anisotropy v_2 from event-by-event calculations could indeed be strongly enhanced from that obtained with average smooth geometry in the same model. A hint for such enhancement was recently reported from the vUSPhydBBMG model [1], which simulates jet energy loss based upon simple parameterized polynomial dependence on path length, medium temperature, and parton energy on top of an event-by-event hydro background. From CIBJET results in Fig. 10.1, however, no significant difference has been detected between the event-by-event case and the average geometry case for either Glauber or Trento initial conditions. To further investigate this issue, let us focus on v_2 at high p_T and compare a number of models in Fig. 10.2. In addition to the CIBJET and vUSPhydBBMG models, three more models are included for this comparison: (1) the CUJET2 model which has a similar DGLV framework as CIBJET but is based on a perturbative quark–gluon medium with HTL resummation [13]; (2) the CLV+LBT model which uses a higher-twist-formalism-based linearized Boltzmann approach in a perturbative quark–gluon medium with simulations on top of the CLV viscous hydro background [14, 15]; (3) the Zakharov's mQGP model [16], which computes the energy loss in the BDMPS-Z formalism based on a medium that adds magnetic monopoles on top of the usual perturbative quark–gluon sector. In between the CUJET2 or CLV+LBT and the CIBJET or mQGP, the main difference is that the latter two models' medium includes a chromo magnetic component. In between CIBJET and mQGP, the main difference is that the CIBJET has its chromo electric component being gradually suppressed toward lower temperature, while the mQGP has no suppression of the quark/gluon sector and directly adds an estimated monopole density [17, 18]. All models nicely describe the same R_{AA} data, and their results for v_2 would provide the critical test. We note in passing other different models not included in this comparison [19–22].

As seen from Fig. 10.2, despite their significant difference in many aspects, both CIBJET (red) and CLV+LBT (black) models demonstrate very small difference between their respective average-geometry results (dashed curves) and event-by-event (solid curves) results. This observation indicates at a limited role of event-by-event fluctuations in the quantitative evaluation of high p_T anisotropy v_2. In comparison with CMS data, the CLV+LBT (black) and the CUJET2 (green dashed) models, both based on a perturbative medium of QGP with HTL resummation, underpredict the v_2 values. The CIBJET (red) and Zakharov mQGP (magenta) models, both including a strong magnetic component near T_c and thus enhancing late time energy loss, give much larger v_2 than the CUJET2 or CLV+LBT model, with the CIBJET in good agreement with data. This comparative study clearly demonstrates the differentiating power of the high p_T anisotropy observable, and strongly suggests two important points: (1) the event-by-event fluctuations have

Fig. 10.2 A comparison of v_2 at high p_T from different models with CMS data [11, 23] for 40–45% Pb+Pb collisions at 2.76 ATeV and 5.02 ATeV, including: CIBJET (red) with event-by-event (solid) or average geometry (dashed), CLV+LBT (black) with event-by-event (solid) or average geometry (dashed), CUJET2 (dashed green), vUSPhydBBMG (dash-dotted blue) and mQGP (dashed magenta)

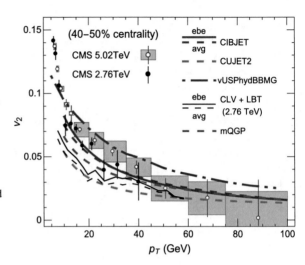

limited impact on the hard sector v_2 values; (2) the inclusion of chromo magnetic component for the medium enhances the hard sector v_2 and is crucial for describing experimental data [3, 24–26].

10.2 Summary of CIBJET Results

In summary, we've established a comprehensive CIBJET framework as a sophisticated and realistic event-by-event simulation tool that allows for a unified, quantitative and consistent description of both soft and hard sector observables $(R_{AA} \otimes v_2 \otimes v_3)$ across a wide span of transverse momentum from ~ 0.5 to ~ 100 GeV and in excellent agreement with experimental data. Such phenomenological success strongly suggests at a highly nontrivial color structure of the near-perfect QCD fluid as a semi-Quark-Gluon-Monopole Plasma (sQGMP), which is in line with the variation of color degrees of freedom as suggested by lattice QCD for the temperature regime most relevant to current heavy ion collision experiments. Remarkably, the sQGMP also provides a dynamical explanation of the temperature-dependent jet transport coefficient $\frac{\hat{q}_F}{T^3}$ and shear viscosity coefficient $\frac{\eta}{s}$ that are mutually consistent as well as consistent with extracted values from phenomenology.

In this study, we show that the CIBJET event-by-event generalization of our previous work does not change our central conclusion that the sQGMP color composition is preferred over the perturbative QCD/HTL composition that is limited to only color screened electric $q(T) + g(T)$ quark and gluon quasi-parton degrees of freedom that are also not consistent with lattice QCD data in the critical crossover temperature range. In this chapter we further show that another composition model [16], referred to as mQGP, that includes magnetic monopoles based on lattice estimations on top of quarks and gluons but does not suppress $q(T) + g(T)$, is also inconsistent with the v_2 data once the coupling is adjusted to reproduce R_{AA}. We conclude this chapter by showing that the sQGMP jet transport coefficient $\hat{q}(T, E)$ peaks near T_c with sufficient strength as to provide a natural dynamical explanation of how the QCD fluid $\eta/s \approx T^3/\hat{q}$ could approach the perfect fluid bound near T_c due to the emergent $m(T)$ component.

References

1. J. Noronha-Hostler, B. Betz, J. Noronha, M. Gyulassy, Phys. Rev. Lett. **116**(25), 252301 (2016). https://doi.org/10.1103/PhysRevLett.116.252301
2. B. Betz, M. Gyulassy, M. Luzum, J. Noronha, J. Noronha-Hostler, I. Portillo, C. Ratti, Phys. Rev. **C95**(4), 044901 (2017). https://doi.org/10.1103/PhysRevC.95.044901
3. J. Xu, J. Liao, M. Gyulassy, Chin. Phys. Lett. **32**(9), 092501 (2015). https://doi.org/10.1088/0256-307X/32/9/092501
4. J. Xu, J. Liao, M. Gyulassy, J. High Energy Phys. **02**, 169 (2016). https://doi.org/10.1007/JHEP02(2016)169
5. S. Shi, J. Xu, J. Liao, M. Gyulassy, Nucl. Phys. **A967**, 648 (2017). https://doi.org/10.1016/j.nuclphysa.2017.06.037
6. S. Shi, J. Liao, M. Gyulassy, Chin. Phys. **C43**(4), 044101 (2019). https://doi.org/10.1088/1674-1137/43/4/044101
7. S. Shi, J. Liao, M. Gyulassy, Chin. Phys. **C42**(10), 104104 (2018). https://doi.org/10.1088/1674-1137/42/10/104104
8. S. Acharya et al., J. High Energy Phys. **11**, 013 (2018). https://doi.org/10.1007/JHEP11(2018)013
9. ATLAS collaboration, ATLAS-CONF-2017-012 (2017)
10. V. Khachatryan et al., J. High Energy Phys. **04**, 039 (2017). https://doi.org/10.1007/JHEP04(2017)039
11. A.M. Sirunyan et al., Phys. Lett. **B776**, 195 (2018). https://doi.org/10.1016/j.physletb.2017.11.041
12. J. Adam et al., Phys. Rev. Lett. **116**(13), 132302 (2016). https://doi.org/10.1103/PhysRevLett.116.132302
13. J. Xu, A. Buzzatti, M. Gyulassy, J. High Energy Phys. **08**, 063 (2014). https://doi.org/10.1007/JHEP08(2014)063
14. S. Cao, L.G. Pang, T. Luo, Y. He, G.Y. Qin, X.N. Wang, Nucl. Part. Phys. Proc. **289–290**, 217 (2017). https://doi.org/10.1016/j.nuclphysbps.2017.05.048
15. S. Cao, T. Luo, G.Y. Qin, X.N. Wang, Phys. Lett. **B777**, 255 (2018). https://doi.org/10.1016/j.physletb.2017.12.023
16. B.G. Zakharov, JETP Lett. **101**(9), 587 (2015). https://doi.org/10.1134/S0021364015090131. [Pisma Zh. Eksp. Teor. Fiz.101,no.9,659(2015)]
17. A. D'Alessandro, M. D'Elia, Nucl. Phys. **B799**, 241 (2008). https://doi.org/10.1016/j.nuclphysb.2008.03.002
18. C. Bonati, M. D'Elia, Nucl. Phys. **B877**, 233 (2013). https://doi.org/10.1016/j.nuclphysb.2013.10.004
19. E. Bianchi, J. Elledge, A. Kumar, A. Majumder, G.Y. Qin, C. Shen (2017). arXiv:1702.00481
20. Y.T. Chien, A. Emerman, Z.B. Kang, G. Ovanesyan, I. Vitev, Phys. Rev. **D93**(7), 074030 (2016). https://doi.org/10.1103/PhysRevD.93.074030

21. M. Djordjevic, M. Djordjevic, B. Blagojevic, Phys. Lett. **B737**, 298 (2014). https://doi.org/10.1016/j.physletb.2014.08.063
22. A. Beraudo et al., Nucl. Phys. **A979**, 21 (2018). https://doi.org/10.1016/j.nuclphysa.2018.09.002
23. S. Chatrchyan et al., Phys. Rev. Lett. **109**, 022301 (2012). https://doi.org/10.1103/PhysRevLett.109.022301
24. X. Zhang, J. Liao, Phys. Rev. **C89**(1), 014907 (2014). https://doi.org/10.1103/PhysRevC.89.014907
25. J. Liao, E. Shuryak, Phys. Rev. Lett. **102**, 202302 (2009). https://doi.org/10.1103/PhysRevLett.102.202302
26. A. Ramamurti, E. Shuryak, Phys. Rev. **D97**(1), 016010 (2018). https://doi.org/10.1103/PhysRevD.97.016010

Chapter 11
Conclusions

The relativistic heavy-ion collision experiments create a new phase of matter, consisting of color-deconfined quarks and gluons, and provide a special environment to study the strong interaction. In the meanwhile, in such collisions produce the hottest mater ($T \sim 10^{12}$ K), the strongest magnetic field ($B \sim 10^{16}$ T), the most vortical system ($\omega \sim 10^{21}$ Hz), and likely the most perfect fluid ($\eta/s \sim 0.1$) in the observed universe. In this dissertation, we take advantage of such extreme environment, and focus on both soft (with transverse momentum $p_T < 2$ GeV) and hard ($p_T > 10$ GeV) probes of the color-deconfined QCD Plasma, especially its topological properties. To this aim, we performed quantitative study of the Chiral Magnetic Effect (CME) which reflects the topological charge transition, as well as jet quenching observables as the probe of the chromo-magnetic degree of freedom.

Firstly, in Chaps. 2–5, we quantitatively studied the Chiral Magnetic Effect in relativistic heavy-ion collisions. CME is a macroscopic manifestation of fundamental chiral anomaly in a many-body system of chiral fermions, and emerges as anomalous transport current in the fluid dynamics framework. We developed the Anomalous-Viscous Fluid Dynamics (AVFD) framework, which implements the anomalous fluid dynamics to describe the evolution of fermion currents in QGP, on top of the neutral bulk background described by the (VISH2+1/MUSIC2+1) hydrodynamic simulations for heavy-ion collisions. With this new tool, we systematically investigated the dependence of the CME signal to a series of theoretical inputs and associated uncertainties, including the time dependence of the magnetic field, the initial axial charge, the viscous transport coefficients, the resonance decay contributions, the possible thermal relaxation effect, as well as the contribution from nontrivial charge or current initial conditions due to the pre-hydro CME. With realistic estimations of initial conditions and magnetic field lifetime, the predicted CME signal is quantitatively consistent with measured change separation data in 200 GeV Au–Au collisions. Finally, based on analysis of Au–Au collisions, we further make predictions for the CME observable to be measured in the ongoing isobaric (Ru–Ru v.s. Zr–Zr) collision experiment. We investigated the influence

© Springer Nature Switzerland AG 2019
S. Shi, *Soft and Hard Probes of QCD Topological Structures
in Relativistic Heavy-Ion Collisions*, Springer Theses,
https://doi.org/10.1007/978-3-030-25482-7_11

of event-by-event fluctuations. In particular, we suggest that comparison between observables within sufficiently narrow $N_{ch} \otimes q_2$ bins would provide a highly decisive test of CME signal in the IsoBar program.

Then we extended the study by considering the chiral effects in out-of-equilibrium stage and/or coupled with vorticity field. First we performed a conceptual study on how one should quantify CME in pre-equilibrium stage in Chap. 6. By deriving the complete and consistent quantum transport theory for massless fermions by applying quantum expansion on Wigner equations, we clarifies the Lorentz invariance and frame dependence issues associated with the quantum correction. Then in Chap. 7 we studied the rotational profile of the QCD fluid. We've used a transport model to investigate the fluid vorticity structures as well as the global Λ hyperon polarization effect, in Au–Au, Cu–Au, and Cu–Cu collisions. A detailed picture of the vorticity structures in those systems was obtained and found to be very similar at the same beam energy and centrality class. We proposed the global hyperon polarization measurements in Cu–Cu and Cu–Au collisions as an ideal and independent verification for the subatomic swirl discovery and for our current interpretation of the observed signal as rotational polarization effect. With future development of hydrodynamic frameworks incorporating both the 3D fluid vorticity structures and the anomalous transport currents, precise experimental measurements on Chiral Vortical Effect signals could provide independent test on the vorticity profile and the topological charge transition in relativistic heavy-ion collisions.

In Chaps. 8, 9, and 10 we employed the jet-quenching measurements to probe the constituents of the QCD plasma formed in heavy-ion collisions. We performed a comprehensive global χ^2 analysis of nuclear collision jet quenching observables from RHIC (200 AGeV), LHC1 (2.76 ATeV), and recent LHC2 (5.02 ATeV) energies. We used the updated CUJET3.1 framework to evaluate jet energy loss distributions in various models of the color structure of the QCD fluids produced in such reactions. The framework combines consistently event-averaged viscous hydrodynamic fields predicted by VISHNU2+1 (validated with soft $p_T < 2$ GeV bulk observables) and the DGLV theory of jet elastic and inelastic energy loss generalized to sQGMP fluids with color structure including effective semi-QGP color electric monopole densities as well as an effective color magnetic monopole density peaking near T_c. We varied the two control parameters of the model: the maximum value of the running QCD coupling, α_c, and the ratio of screening masses, $c_m = \mu_M/\mu_E$, and made predictions for the $p_T > 10$ GeV, the centrality, as well as the $\sqrt{s} = 0.2$–5.02 ATeV dependence of the nuclear modified jet fragment observables R_{AA} and v_2. The global χ^2 is found to be minimized to near unity for $\alpha_c \approx 0.9 \pm 0.1$, and $c_m \approx 0.25 \pm 0.03$. Thus, with chromo-magnetic monopoles, CUJET3.1 provides a non-perturbative solution to the long standing hard (R_{AA} and v_2) versus soft "perfect fluidity" puzzle. Predictions for future tests at LHC with 5.44 ATeV Xe–Xe and 5.02 ATeV Pb–Pb are also presented. We further developed the CIBJET framework, which combines consistently predictions of event-by-event VISHNU2+1 viscous hydrodynamic fluid fields with CUJET3.1 predictions of event-by-event jet quenching. While being essential to the non-vanishing v_3, event-by-event fluctuations of the bulk background is found to have only limited influence on v_2, and do not affect the direction averaged nuclear modification factor R_{AA}.

To summarize with key messages, we performed systematic quantitative study on CME signal in 200 GeV Au–Au collisions, and the obtained CME signal is quantitatively consistent with the measured charge separation data. We further made predictions for the CME observable in the isobaric (Ru–Ru v.s. Zr–Zr) collisions, and expected such contrast experiments could provide the most decisive test of the CME in heavy-ion collisions. Then we found that recent jet-quenching data favor a temperature dependent color composition including bleached chromo-electric components and an emergent chromo-magnetic degrees of freedom consistent with non-perturbative lattice QCD information in the confinement/deconfinement temperature range.

Curriculum Vitae

Shuzhe Shi (施舒哲)

Department of Physics, McGill University,
3600 University Street, Montreal,
Quebec H3A 2T8, Canada.

e-mails: *shuzhe.shi@mcgill.ca*
shuzhe.shi@gmail.com

▬▬▬ Professional Experience

2018.10 **PostDoc Fellow**, McGill University, Canada.
– present Advisors: Drs. Charles Gale & Sangyong Jeon

2016.06 **Research Assistant**, Indiana University, USA.
– 2018.07 Advisor: Dr. Jinfeng Liao

2012.09 **Research Assistant**, Tsinghua University (清华大学), China.
– 2015.07 Advisor: Dr. Pengfei Zhuang

▬▬▬ Education

2015.08 **Ph.D.**, *Dept. of Physics*, Indiana University, Bloomington, IN, USA.
– 2018.07 - Supervisor: Dr. Jinfeng Liao (廖劲峰)
- Thesis: Soft and Hard Probes of QCD Topological Structures in Relativistic Heavy-Ion Collisions

2012.09 **M.S.**, *Dept. of Physics*, Tsinghua Univ. (清华大学), Beijing, China.
– 2015.07 - Supervisor: Dr. Pengfei Zhuang (庄鹏飞)
- Thesis: Fluctuations and Quarkonia in High Energy Nuclear Collisions

2008.09 **B.S.**, *Dept. of Physics*, Tsinghua Univ. (清华大学), Beijing, China.
– 2012.07 - Supervisor: Dr. Pengfei Zhuang (庄鹏飞)
- Thesis: Correlations between Collective Flow Fluctuations and Thermal Fluctuations in Relativistic Expanding System

© Springer Nature Switzerland AG 2019
S. Shi, *Soft and Hard Probes of QCD Topological Structures in Relativistic Heavy-Ion Collisions*, Springer Theses,
https://doi.org/10.1007/978-3-030-25482-7

━━━━ # Honors

2017.04   Outstanding Graduate Student Award
          Department of Physics, Indiana University
          - *In recognition of performance as an Outstanding Graduate Student in Research.*

2015.07   Distinguished Thesis, Master (优秀硕士学位论文)
          Tsinghua University
          - *For selected ~200 best out of ~4000 theses of all master programs.*

2015.07   Distinguished Graduate, Master (优秀硕士毕业生)
          Tsinghua University
          - *Recognizing ~100/4000 excellent master graduates of all programs.*

2015.07   Yu-Hsun Woo Prize (吴有训奖)
          Dept. of Physics, Tsinghua Univ. & Hung Yin Hua Guan Foundation
          - *For excellent graduate students majoring in physics.*

2013.12   National Scholarship for Master Students (硕士研究生国家奖学金)
          Ministry of Education of China (中华人民共和国教育部)
          - *Scholarship for outstanding master students.*

2012.05   Tsinghua Xuetang Scholarship (清华学堂奖学金), Twice
&2011.12 Tsinghua University
          - *Scholarship provided by the "Tsinghua Xuetang Talents Program" to support outstanding undergraduate students majoring in fundamental science.*

━━━━ # Publications

━━━ ## In Peer-Reviewed Journal

16)   **S. Shi**, J. Liao and M. Gyulassy, *"Global Constraints from RHIC and LHC on Transport Properties of QCD Fluid with the CUJET/CIBJET Framework"*, Chin. Phys. C 43 (2019) 4, 044101, arXiv:1808.05461 [hep-ph].

15)   N. Magdy, **S. Shi** J. Liao, P. Liu and R. A. Lacey, *"Examination of the observability of a chiral magnetically-driven charge-separation difference in collisions of the $^{96}_{44}Ru + ^{96}_{44}Ru$ and $^{96}_{40}Zr + ^{96}_{40}Zr$ isobars at energies available at RHIC,"*, Phys. Rev. C 98, 061902 (2018), arXiv:1803.02416 [nucl-ex].

14)   **S. Shi**, K. Li, J. Liao, *"Searching for the Subatomic Swirls in the CuCu and CuAu Collisions "*, Phys. Lett. B 788 (2019) 409, arXiv:1712.00878 [nucl-th].

13) **S. Shi**, J. Liao and M. Gyulassy, *"Probing the Color Structure of the Perfect QCD Fluids via Soft-Hard-Event-by-Event Azimuthal Correlations"*, Chin. Phys. C 42 (2018) 10, 104104, arXiv:1804.01915 [hep-ph].

12) R. Rapp, P.B. Gossiaux, A. Andronic, R. Averbeck, S. Masciocchi, A. Beraudo, E. Bratkovskaya, P. Braun-Munzinger, S. Cao, A. Dainese, S.K. Das, M. Djordjevic, V. Greco, M. He, H. van Hees, G. Inghirami, O. Kaczmarek, Y.-J. Lee, J. Liao, S.Y.F. Liu, G. Moore, M. Nahrgang, J. Pawlowski, P. Petreczky, S. Plumari, F. Prino, **S. Shi**, T. Song, J. Stachel, I. Vitev, X.-N. Wang, *"Extraction of Heavy-Flavor Transport Coefficients in QCD Matter"*, Nucl. Phys. A 979 (2018) 21-86, arXiv:1803.03824 [nucl-th].

11) A. Huang, **S. Shi**, Y. Jiang, J. Liao and P. Zhuang, *"Complete and Consistent Chiral Transport from Wigner Function Formalism"*, Phys. Rev. D 98 (2018) 036010, arXiv:1801.03640 [hep-th].

10) N. Magdy, **S. Shi**, J. Liao, N. Ajitanand and R. A. Lacey, *"A New Correlator to Detect and Characterize the Chiral Magnetic Effect"*, Phys. Rev. C 97 (2018) 061901(R), arXiv:1710.01717 [physics.data-an].

9) B. Feng, C. Greiner, **S. Shi**, Z. Xu, *"Viscous effects on QCD first-order confinement phase transition"*, Phys. Lett. B 782 (2018) 262, arXiv:1802.02494 [hep-th].

8) **S. Shi**, Y. Jiang, E. Lilleskov and J. Liao, *"Anomalous Chiral Transport in Heavy Ion Collisions from Anomalous-Viscous Fluid Dynamics"*, Annals Phys. 349 (2018) 50-72, arXiv:1711.02496 [nucl-th].

7) A. Huang, Y. Jiang, **S. Shi**, J. Liao and P. Zhuang, *"Out-of-Equilibrium Chiral Magnetic Effect from Chiral Kinetic Theory"*, Phys. Lett. B 777 (2018) 177, arXiv:1703.08856 [hep-ph].

6) Y. Jiang, **S. Shi**, Y. Yin and J. Liao, *"Quantifying Chiral Magnetic Effect from Anomalous-Viscous Fluid Dynamics"*, Chin. Phys. C 42 (2018) 1, 011001, arXiv:1611.04586 [nucl-th].

5) X. Guo, **S. Shi**, N. Xu, Z. Xu, P. Zhuang, *"Magnetic Field Effect on Charmonium Production in High Energy Nuclear Collisions"*, Phys. Lett. B 751 (2015) 215, arXiv:1502.04407 [hep-ph].

4) **S.Shi**, J. Liao, P. Zhuang, *"'Ripples' on Relativistic Expanding Fluid"*, Phys. Rev. C 90 (2014), 064912, arXiv:1405.4546[hep-th].

3) **S. Shi**, X. Guo, P. Zhuang, *"Flavor Dependence of Meson Melting Temperature in Relativistic Potential Model"*, Phys. Rev. D 88 (2013) 1, 014021, arXiv:1306.7752 [nucl-th].

2) **S. Shi**, J. Liao, *"Conserve Charge Fluctuations and Suscepti-bilities in Strongly Interacting Matter"*, JHEP 1306 (2013) 104, arXiv:1304.7752 [hep-ph].

1) X. Guo, **S. Shi**, P. Zhuang, *"Relativistic Correction to Char-monium Dissociation Temperature"*, Phys. Lett. B 718 (2012) 143, arXiv:1209.5873 [hep-ph].

Submitted

4) Y. Guo, **S. Shi**, S. Feng, and J. Liao, *"Magnetic Field In-duced Polarization Difference between Hyperons and Anti-hyperons"*, arXiv:arXiv:1905.12631 [nucl-th].

3) **S. Shi**, J. Zhao and P. Zhuang, *"Heavy Flavor Hadrons from Multi-Body Dirac Equations"*, arXiv:arXiv:1905.10627[nucl-th].

2) Z. Wang, X. Guo, **S. Shi** and P. Zhuang, *"Mass Correction to Chiral Kinetic Equations"*, arXiv:1903.03461 [hep-ph].

1) J. Zhao, **S. Shi**, N. Xu and P. Zhuang, *"Sequential Coales-cence with Charm Conservation in High Energy Nuclear Collisions"*, arXiv:1805.10858 [hep-ph].

In Refereed Conference Proceeding

6) J. Zhao, **S. Shi**, N. Xu and P. Zhuang, *"Conservational and Sequential Charm Hadronization in Heavy Ion Collisions"*, EPJ Web Conf. 202 (2019) 06004. Contribution to 9th Interna-tional Workshop on Charm Physics (CHARM 2018).

5) **S. Shi**, H. Zhang, D, Hou, and J. Liao, *"Chiral Magnetic Effect in Iso-baric Collisions from Anomalous-Viscous Fluid Dynamics (AVFD)"*, Nucl.Phys.A982(2019)539-542, arXiv:1807.05604 [hep-ph]. Contri-bution to the Proceedings of XXVIIth International Conference on Ultrarelativistic Nucleus-Nucleus Collisions (Quark Matter 2018).

4) M. Gyulassy, P. Levai, J. Liao, **S. Shi**, F. Yuan, X.N. Wang, *"Precision Dijet Acoplanarity Tomography of the Chromo Struc-ture of Perfect QCD Fluids"*, Nucl. Phys. A 982 (2019) 627-630, arXiv:1808.03238 [hep-ph]. Contribution to the Proceedings of XXVI-Ith International Conference on Ultrarelativistic Nucleus-Nucleus Col-lisions (Quark Matter 2018).

3) **S. Shi**, Y. Jiang, E. Lilleskov, Y. Yin, J. Liao, *"Chiral Magnetic Effect from Event-by-Event Anomalous-Viscous Fluid Mechanics"*, PoS CPOD2017 (2018) 021, arXiv:1712.01386 [nucl-th]. Contribu-tion to the Proceedings of "CPOD2017: Critical Point and Onset of Deconfinement" Conference.

2) **S. Shi**, Y. Jiang, E. Lilleskov, Y. Yin, J. Liao, *"Quantifying the Chiral Magnetic Effect from Anomalous-Viscous Fluid Dynamics"*, Nucl. Phys. A 967 (2017) 748-751,   arXiv:1704.05531 [nucl-th]. Contribution to the Proceedings of XXVIth International Conference on Ultrarelativistic Nucleus-Nucleus Collisions (Quark Matter 2017).

1) **S. Shi**, J. Xu, J. Liao and M. Gyulassy, *"A Unified Description for Comprehensive Sets of Jet Energy Loss Observables with CUJET3"*, Nucl. Phys. A 967 (2017) 648-651,   arXiv:1704.04577 [hep-ph]. Contribution to the Proceedings of XXVIth International Conference on Ultrarelativistic Nucleus-Nucleus Collisions (Quark Matter 2017).

▬▬ Presentations

▬ Invited Talks

- *"Chiral Magnetic Effect from Event-by-Event Anomalous-Viscous Fluid Dynamics"*, Invited **plenary** talk at Critical Points and Onset of Deconfinement 2017, Stony Brook University, Stony Brook, USA (Aug. 2017).

- *"Anomalous-Viscous Fluid Dynamics"*, 5th edition of the Workshop on Chirality, Vorticity and Magnetic Field in Heavy Ion Collisions, Tsinghua University, Beijing, China (Apr. 2019).

- *"Probing the Constituent of the QCD Plasma via CUJET3.1/CIBJET Framework"*, 13th International Workshop on High-pT Physics in the RHIC/LHC Era, University of Tennessee, Knoxville, USA (Mar. 2019).

- *"Probing the QCD Plasma Constituent via CUJET3.1/CIBJET Framework"*, RBRC Workshop on the Definition of Jets in a Large Background, BNL, Brookhaven, USA (Jun. 2018).

- *"Quantitative Study of Chiral Magnetic Effect from Anomalous-Viscous Fluid Dynamics"*, Workshop on Chirality, Criticality and Correlations in Heavy Ion Collisions, Sun Yat-Sen University, Guangzhou, China (Nov. 2017).

- *"Quantifying CME in Isobaric collisions"*, 2017 RHIC & AGS Annual User's Meeting, BNL, Brookhaven, USA (Jun. 2017).

- *"Conserved Charge Transport within the AMPT Model"*, QCD Chirality Workshop 2016, UCLA, Los Angeles, USA (Feb. 2016).

▬ Contributed Talks

- *"Anomalous-Viscous Fluid Dynamics"*, Topical Workshop on Beam Energy Scan (BEST 2017), Stony Brook University, Stony Brook, USA (Aug. 2017).

- *"Quantifying Chiral Magnetic Effect from Anomalous-Viscous Fluid Dynamics"*, Quark Matter 2017, Chicago, USA (Feb. 2017).
- *"CME from Chiral Viscous Hydrodynamics"*, Topical Workshop on Beam Energy Scan (BEST 2016), Indiana University, Bloomington, USA (May. 2016).
- *"Quarkonia in Strong Magnetic Field"*, Workshop of QCD Vacuum and Matter under Strong Magnetic Field, IHEP, Beijing, China (Oct. 2014).
- *"Flavor Dependence of Meson Melting Temperature in Relativistic Potential Model"*, the 15th National Conference on Nuclear Physics (第十五届全国核物理大会), Shanghai, China (Oct. 2013).

───── **Seminars**

- *Quantitative Study of Chiral Magnetic Effect from (Event-by-Event) Anomalous-Viscous Fluid Dynamics*:

    – Institute of Quantum Matter, South China Normal Univ., Guangzhou, China (Apr. 2019).

    – Dept. of Physics, Tsinghua Univ., Beijing, China (Nov. 2017).

    – Physics Dept., Ohio State Univ., Columbus, USA (Apr. 2017).

    – Dept. of Physics, Tsinghua Univ., Beijing, China (Dec. 2016).

- *Soft and Hard Probes of QCD Topological Structures in Relativistic Heavy-Ion Collisions*:

    – Physics Dept., McGill Univ., Montreal, Canada (Nov. 2018).

- *Probing the Constituent of the QCD Plasma via CUJET3.1/CIBJET Framework*:

    – Dept. of Physics, Tsinghua Univ., Beijing, China (Jul. 2018).

- *Heavy Flavor Hadron Mass Spectrum from Three-Body Dirac Equations*:

    – Dept. of Physics, Tsinghua Univ., Beijing, China (Apr. 2019).

Printed in the United States
By Bookmasters